AN IDENTIFICATION GUIDE TO

GARDEN INSECTS OF BRITAIN

and North-West Europe

Dominic Couzens
and Gail Ashton

JOHN BEAUFOY PUBLISHING

First published in the United Kingdom in 2022 by John Beaufoy Publishing Ltd
11 Blenheim Court, 316 Woodstock Road, Oxford OX2 7NS, England
www.johnbeaufoy.com

Photo Credits
Front cover: Twenty-two-spot Ladybird © Gail Ashton
Back cover: *top* Common Carder Bumblebee, *middle* Hawthorn Shieldbug, *bottom* Marmalade Hoverfly, all © Gail Ashton
Spine: Lily Beetle © Gail Ashton
Dominic Couzens p. 126 (bottom left); Bob Gibbons pp. 40 (bottom right); 46, 50 and 118.

Consultants: Dr Bob Gibbons, Lucy Boulton, Dr Ross Piper

ISBN 978-1-913679-25-5

Edited by Krystyna Mayer
Designed by Nigel Partridge
Project Management by Rosemary Wilkinson
Printed and bound in Malaysia by Times Offset (M) Sdn. Bhd.

CONTENTS

INTRODUCTION

This book has been written to help with recognition of the insects you might find in your garden or nearby green space and – more importantly – to encourage you to appreciate them. In writing it, we want to share our admiration and fascination for insects so that you might enjoy these diminutive neighbours more and more.

LOVE INSECTS?

Perhaps the first question you might have is why would you even want insects in your garden in the first place? Well, do you like birds, hedgehogs, badgers, bats and foxes? If the answer is yes to any of the above, then you need lots of insects. Basically, insects are food for pretty much everything. Without them, almost everything higher up the food chain would starve, including humans. Obviously, the majority of us do not eat insects (yet), but without bees, wasps, flies, beetles and even earwigs, most of our fruits and vegetables would not exist. Believe it or not, we need insects a lot more than we know.

Insects are in decline; they are in danger of dying out faster than they can reproduce. The main reason for this decline is habitat loss. You may think that having fewer wasps and midges bothering you in your summer garden sounds great, but without them many other animals would suffer.

Imagine, for a moment, that your closest food shop is over 100 kilometres away and you have to walk to it. You arrive there, exhausted and hungry, only to find that the shop shelves are almost empty. If this sounds like a unlikely scenario, it is becoming an increasingly common reality for migratory birds such as swallows and swifts, which eat mosquitoes, midges and other small flying insects. They fly thousands of kilometres to feed on ever-dwindling food sources, and a decrease in insects would result in a decrease in these extraordinary summer visitors. This is why it is so important to protect and nurture our insect communities.

ATTRACTING INSECTS TO YOUR GARDEN

Attracting insects into your garden is actually really easy. Follow these simple steps and you will see an increase in biodiversity in no time.

First of all, here is what you *should not* do:

1. Don't use plastic grass Astro turf is fine for hockey pitches, but it is the worst thing for your garden. Soil-dwelling animals such as worms, moles and ground beetles get trapped underneath it, so that birds such as robins and blackbirds lose a valuable food source. Plastic grass does not grow flowers either – which is bad news for pollinating insects. The other terrible thing about plastic grass is that when it wears out, it ends up in landfill forever. Real grass regenerates itself and is completely environmentally friendly.

2. Don't use chemicals Never use pesticides. They harm *all* animals, including us. Think of all the invertebrates that you have been told are 'bad', such as aphids, wasps, slugs and spiders. These animals are essential food for birds. We all love to feed our garden birds and spend a lot of money each year on bird food such as suet balls and peanuts. What if we stopped using pesticides instead, and let these 'pests' run riot in our gardens? Well, they would get eaten by birds. More invertebrates = more birds = fewer pests. Many insects are carnivorous and eat other insects that damage plants. Birds, mammals and insects all provide much better natural pest control than chemicals.

3. Don't mow a lawn too often It is good to have a grass lawn, and infinitely better than having a plastic one. However, many people keep the sward short with regular mowings, which as far as insects are concerned is not ideal since it limits the number of glorious rich herbs that pollinators favour; bees, flies and moths love daisies, dandelions and clovers. While a short sward looks neat, and is of course a matter of personal choice, why not join the growing band of people who mow their lawns only occasionally, perhaps only in midsummer and again in autumn? Avoid mowing between May and late August. Even leaving small sections unmown will make a big difference.

4. Don't use peat compost Always avoid compost with peat in it. It is not so much about your garden, but about where the peat is sourced, which is often a wild peat bog that is a wonderful habitat and traps a lot of carbon. It will be banned eventually, but until then try to source peat-free compost or even make your own.

Making a Garden into an Oasis for Wildlife

This is what you can do:

1. Relax Let the grass and hedges grow a little longer, and do not obsess about the garden being too tidy. The best way to create a habitat is to not do much at all.

2. Rewild By rewilding you can stop managing your garden altogether and let it grow rampant. In reality, this is not always an option, but everyone can rewild a part of their garden, even a tiny corner. It will make such a difference. Choose a sunny patch and let it do its own thing. See what grows – it could be nettles and brambles, or wildflowers. It will quickly become a haven for bugs, beetles and caterpillars, as well as spiders, woodlice and even wood mice. Put some logs in there – beetles and fungi will love them, and you may even get a hedgehog hibernating in the logs.

3. Create a wildlife corridor According to the Office of National Statistics, around half a million hectares of urbanized land is made up of residential gardens. If your garden is fully fenced, you could cut a small hole at the bottom of it to create a wildlife highway for frogs, toads, hedgehogs and small rodents. Why not plant a hedge, too – if not in your garden, perhaps in your neighbourhood? Joined-up habitats are better than isolated ones.

4. Make a pond If you do one thing to help wildlife in your garden, make a pond. You can use a washing-up bowl or a bathtub, or create a lake if you want – it really does not matter. Water sources are essential to life, and gardens become all the richer for ponds. Locate a pond in a partly sunny place, with lots of vegetation around it where insects can shelter. Within a couple of weeks you will have hoverflies investigating it, and possibly laying eggs in it. Within a year or two you may have water beetles and damselfly larvae, as well as frogs or newts. Ponds also provide a critical water supply for birds, hedgehogs, foxes and other wildlife.

5. Plant a fruit tree Insects and birds love shelter. Hedges and trees are their favourite places to sleep in and hide from predators. Planting an apple or pear tree in your garden will be a long-term gift to your garden wildlife and will also provide you with some free fruit in the autumn.

6. Choose the best plants for your garden We have all heard that pollinators need flowers, and many people are keen to put in flowers that will attract bees. However, 'bee-friendly' labels at garden centres are misleading. If you must shop at a garden centre or nursery, go on a warm day, take your time and observe for yourself which flowers bees are visiting. Also try to source flowers from your friends and neighbours, and include some native species in your planting scheme.

7. Be patient You will not get all of the insects described in this book in your garden, and some you may never see. However, it can be guaranteed that, by creating some of the habitats mentioned above, many will appear. Some insects, such as hoverflies, will appear within days if you build a pond and have carefully chosen summer flowering plants (see above). Others, such as solitary bees and moths, may appear the following spring. It takes a long time to build a fully functioning ecosystem, so it is perfectly acceptable to allow your garden time to become naturally attractive to wildlife over a period of months and even years.

THE INSECT LOVER'S YEAR

Insects can be found at all times of the year – you just have to know where to look. In winter and early spring, you may find hibernating ladybirds in your shed or in gaps around windowsills. Take a walk in the woods and lift some logs; at some point you will come across a sleepy ground beetle or earwig. There are even a few moths that come out during winter. But generally the only flying insects you will see during the colder months are large bumblebee queens, which occasionally emerge to refuel on nectar from winter-flowering plants. The dense, chitinous fuzz that bumblebee queens possess allows them to heat their bodies and flight muscles in temperatures as low as 5°C.

Late March is traditionally the time that the entomological starting gun goes off. Those bumblebee queens will inevitably be among the first flying insects you will see in early spring, but on sunny days the first solitary bees will start to emerge, along with that superstar fly of early season, the fuzzy, long-snouted Dark-edged Bee-fly. They are followed by the hardy Hairy-footed Flower Bees in mid-March, which take advantage of the lack of traffic around spring-flowering plants such as lungwort and grape hyacinth, taking their fill until early April, when temperatures rise to the mid-teens and the bee bun-fight really begins.

The sun starts to warm the air, and it fills with the sound of buzzing. If you have a good range of pollinator-friendly flowers in your garden, it will become a hive of activity. Late spring and early summer are peak insect season, when the beetles and solitary wasps all emerge to mate, and feed on the plentiful nectar and pollen available. These are the months that insect enthusiasts look forward to the most. There is that first day in spring, when you realize that the still, silent air around you suddenly resonates with buzzing and zigzagging movements. The dawn chorus of birds is a marvellous sound, but for some nothing sounds quite like the hum of insect wings darting through the warm sunlight.

If you have a good variety of flowers and habitats in your garden, then through the warm summer months you may be visited by all manner of hoverflies, bees, small wasps, bugs, beetles, ants, butterflies and moths and, if you have water nearby, damselflies and dragonflies. Even the nights are filled with insects. The moths are some of the most beautiful and striking members of our wildlife community, yet are rarely seen because the majority of them are nocturnal. To see them, go into your garden after dark with a torch on a warm, windless, slightly cloudy night. You will see moths of all shapes, sizes and colours flying around, or nectaring on flowers.

The warm summer months are lovely for us, but they are critical for insects. This is the time when most insects emerge in their adult stage with the sole purpose of reproducing. The majority of them will die by the time autumn sets in, so they have just a few months in which to emerge, fuel up and find a mate. If you have a really good look around, in masonry holes, under leaves and on the stems of plants, you may well find tiny eggs or nest cells. Some of these will hatch within a few days or weeks, while others will spend the entire autumn and winter developing, and emerge the following spring. If you have long grass anywhere, check for bush-crickets, grasshoppers and plant bugs.

The peak invertebrate season is fleeting, from April to late June. Why is it so brief? Well, the adult stages of most insects are relatively short. For example, the Stag Beetle lives out most of its lifespan underground, spending an astonishing 5–7 years as a fat, squidgy larva, chewing on dead wood. Its eventual metamorphosis into the above-ground adult we recognize is actually its swansong. It emerges in late May or early June with a shelf life of just a few months at best to seek out a mate and secure the next generation, literally its only job before its death in late summer. This may seem utterly crazy. Half a decade underground and only weeks as our largest native insect? Well, evolution is crazy, and humans have a propensity to lay more importance at the feet of things we can see and hear in front of us. The fact that Stag Beetle larvae are an essential decomposer and the shiny, photogenic adults are merely a reproductive by-product can be lost on us.

By the end of summer and early autumn, the vast majority of garden insects will have disappeared, either reaching the end of their life cycle, or hibernating. The level of activity will ebb away until eventually all becomes quiet. However, this does not mean that there is nothing left to discover. Some species can be found well into early autumn. Look under a log or plant pot on a cold winter day and you may find a snoozing ground beetle (as well as the ubiquitous woodlice and slugs).

By mid-autumn, the number of insects really does decline. However, the fuse for the next season is already lit. Check the cracks in tree bark in early January and you may happen upon a ladybird, tucked up and sheltering from the cold. In the eaves of your balcony, or in a garage, a Small Tortoiseshell butterfly may be sleeping off winter. Just under the surface of the soil, there will be beetle larvae, bee pupae and even bumblebee queens, patiently awaiting spring.

GOING ON A GARDEN SAFARI

Whatever type of garden you have, insects will never be far away. However, because they are mostly very small, you may have to try a few different methods to find them. Growing flowers is an obvious tactic. The mere presence of lavender, whether a large clump in a garden, or a single plant in a container on a balcony, will attract several species of bee in the summer months. A winter-flowering honeysuckle or mahonia will, amazingly, attract bumblebee queens in January. The more pollinator-friendly plants you have, the more insects you will see – it is that simple.

And once you start to attract insects, how do you identify and record them? Some insects prefer to stay hidden in vegetation, others favour areas around the bases of plants on or in the soil, and some the tree canopy. Many only come out at night, when we are asleep. However, there are several, very easy ways in which you can find your more elusive garden residents.

Tapping This is an excellent way to find out what is hiding in your trees and shrubs. All you need is an old, light-coloured bedsheet and a good-sized stick. Spread out the sheet on the ground, underneath a tree. Then give the branches and leaves a firm but careful tap all over. You can also shake the branches. You should see a variety of beasts drop on to the sheet. Equip yourself with a magnifying glass or hand lens, so that you can get a close-up look at the wings and patterns. Winged insects will fly back up into the canopy, but non-flyers should be placed carefully back on leaves and branches.

Moth trap Some of our most colourful and charismatic insects are crepuscular (twilight-loving) and nocturnal, such as moths, some beetles and some ichneumon wasps. Professional moth traps are expensive and require careful handling, but have a look online – there are some great video tutorials for DIY versions that can be made using LEDs. An even simpler method is to again use a light-coloured bedsheet (or similar), hang it up nice and straight on a wall or fence at dusk, and shine a bright torch at it. Be patient; you may start to see moths fly to the light and land on the sheet – then you can have a close look at them or photograph them. Different moths fly at different times of the night – the longer you stay up, the more species you could see.

Pitfall trap A pitfall trap provides a great way to observe ground insects, particularly beetles. For making this, a large yogurt pot is perfect. Dig a hole in the soil and sink the pot into the ground, making sure that the rim is level with the soil. Check the trap over the next few days, to see if any interesting insects have fallen into it. Remember to release any insects that you find, and do not abandon the trap; take it out of the ground once your monitoring period is over, so that insects do not get stuck in it.

First, try to identify anything you have found. This book will help you to identify a lot of the common insects, or at least to work out what family they are in. Record your findings. Take photographs of things you find, and post them to one of the many recording schemes that are being run online (see p. 158). Even if you do not know exactly which insect you have, there are lots of people who will help you identify your garden insects. You can also be a critical part of monitoring our insect populations.

PHOTOGRAPHING INSECTS

Insect photography has become hugely popular in recent years. More people are seeing just how beautiful insects are close up, and are keen to find and record this hugely diverse group of animals. There are more than 24,000 described species of insect in the UK, compared to around 600 bird species and 122 mammal species. This statistic alone shows just what an exciting prospect it is to study and photograph insects; you might even find a new one. All insects are small compared to most mammals; they are certainly tiny in comparison to humans. It is actually quite difficult to get a good look at them with just our eyes. This is a real shame, because insect exoskeletons contain so much surface detail that we just cannot see. To really explore these magnificent animals you need to get close up. You could use a hand lens or magnifying glass, but insects move around quickly, or hide from us, so it can be tricky to get a clear view. You can also use a microscope, but this usually involves 'collecting a specimen', which means that the insect has to be killed in order to be observed through the microscopic lens. This should only ever be done by professional entomologists, who follow a strict ethical code. So, what is the solution? Taking photographs provides a great way of recording sightings, not only for ourselves, but also for contributing to citizen science projects.

If you have the right equipment, insect photography is by far the easiest form of wildlife photography. You can get much closer to many insects than to, say, birds or mammals. So, what is the right equipment? Macro photography used to be an expensive business, requiring elite cameras, lenses and lighting. Specialists have very high-end kits including a professional D-SLR camera, with a high-quality macro lens. They may also have a sophisticated flash system. Of course, most people cannot or do not want to spend a huge amount of money for this purpose. However, for a few hundred pounds, you can find some fantastic compact cameras with a macro setting that will give you very pleasing results. Smartphone optics continue to improve at a thunderous rate. The top phones have decent macro capability, and you can improve this further with a clip-on or stick-on macro lens that comes at a fraction of the price of a professional camera kit. You can choose to spend thousands of pounds, a few hundred or as little as £20. All of these options will give you a fascinating view into the tiny yet exquisite world of insects.

When photographing insects, you need to follow a few golden rules:

1. Take your time. Many insects do not like sudden movements and will flee at the first sign of a threat. Move slowly towards your subject, keeping all of your movements slow and smooth.

2. Do not shout or talk loudly – sound can disturb insects, too. Move your camera gradually towards an insect, taking shots as you go. Hopefully you will get really close and capture all that amazing detail.

3. Cast no shadow. Winged insects often take flight when your shadow moves over them – they either think you are a predator, or need to move back into the sun to stay warm. Position yourself so that you do not cast a shadow over them by putting them in between yourself and the sun.

4. Do no harm. Do not chase an insect through its habitat – you could disturb other wildlife, or worse still, hurt it or destroy nesting sites. Do not handle an insect unless you are super confident that you will not harm it.

Parts of an insect.

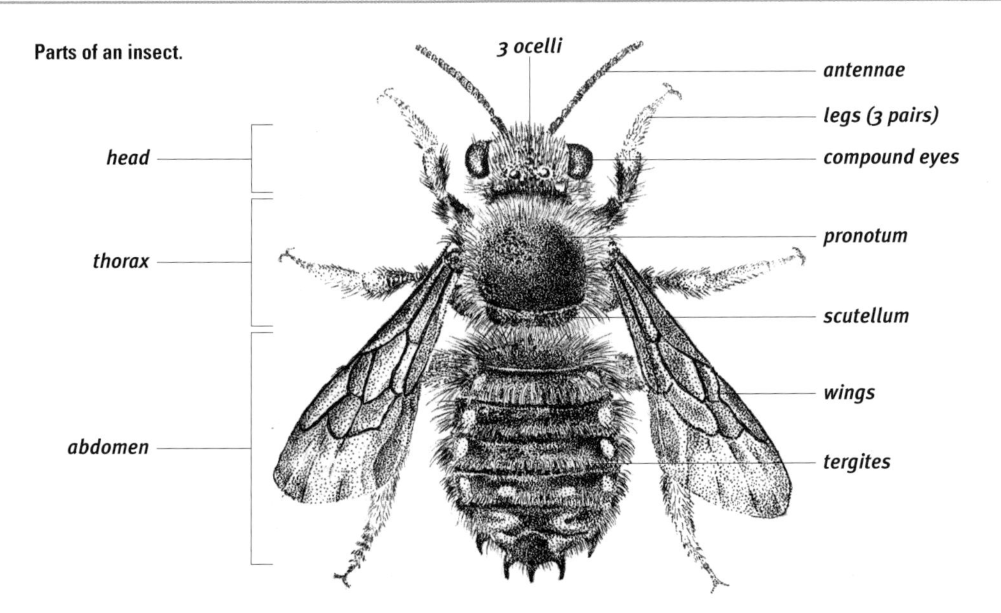

With this information in hand, get out with your cameras and phones to explore and record the extraordinary world of insects. Use this book to help you identify an insect you have photographed. If it is not here, there are lots of other resources to help you. You can also record your sightings on the many biological monitoring apps available. Help is needed to find out what is out there, and your records will be valuable in this respect (p. 158).

WHAT IS AN INSECT?

Insects are a class (a higher grouping) of animals with an exoskeleton (external skeleton) made up from chitin, as well as the following characteristics:

- A body made up of three parts – head, thorax and abdomen.
- Three pairs of jointed legs.
- One pair of antennae.
- Wings – one or two pairs, visible or concealed beneath wing casings.

The last criterion is less rigid; many insect species do not have wings, but they may be found in some life-cycle stages, for example in a newly emerged queen and male. Some ants (known as alates) have wings, whereas worker ants do not. Beetles have hard wing casings and longer antennae. Beetle wings are not visible. They are concealed beneath hardened outer casings called elytra, which are actually modified forewings. The wings are folded intricately inside and can unfold in an instant, ready for take-off. The elytra also protect the beetles' soft, vulnerable bodies from damage. Some beetles, such as the Bloody-nosed Beetle, have fused

The elytra of the Lesser Stag Beetle.

The halteres on St Mark's Fly.

elytra; their wing casings are fixed and therefore they do not fly. Flies have one pair of wings, shorter antennae and reduced, modified hindwings called halteres. Bees have two pairs of wings, medium-slim antennae, and are often furry or hairy. Wasps have two pairs of wings, medium-slim antennae and very slim waists.

Some other parts are indicated in the diagram on p. 9 and these terms are used throughout the book.

Flies are the only order of insects that have halteres. As shown here in a member of the Bibionidae family (March flies) these are the appendages that stick out like tiny wings. They help with balance and turning in flight. They can vary in shape and size.

Damselfly eyes (below left) are separated and sit on either side of the head. Dragonfly eyes (below right) curve around the head and meet at the top.

THE INSECTS IN THIS BOOK

There are a bewildering 24,000 insect species in Britain, and nearly three times that in mainland Europe. Cutting that down to 150 or so species was therefore difficult. However, we have chosen a range of the most common insects that you are likely to come across next to houses, and representatives from as many insect groups as possible. It was not an easy choice.

Due to their amazing diversity, to get a handle on them you need to have some understanding of what insects are and what types there are.

All insects undergo metamorphosis, a change of body form as they grow. There are usually four stages – egg, larva, pupa and adult (also called imago). In some groups, such as dragonflies and true bugs, there is no pupal stage. The immature stages often have a completely different lifestyle

Blue-tailed Damselfly's eyes.

Common Darter's eyes.

from the adults. The main growing stage is the larva, which may moult many times. If the larva is a small version of the adult, it is known as a nymph.

Within the class of insects there is a large number of different orders. Here are the characteristics of the orders included in this book (adapted from Brock, 2019):

EPHEMEROPTERA Mayflies

One or two pairs of transparent wings, with forewings always much larger than hindwings; long, slim body; short antennae; three-pronged 'tail'.

- Incomplete metamorphosis: egg, larva (nymph), adult.

ODONATA Dragonflies & Damselflies

Two pairs of wings allow for powerful, agile flight; large eyes and jaws; almost invisible antennae; elongated, narrow abdomen.

- Incomplete metamorphosis: egg, larva (nymph), adult.

DERMAPTERA Earwigs

Familiar flat-bodied insects with prominent 'pincers' (cerci) at the rear end. Hindwings folded and concealed under forewings.

- Incomplete metamorphosis: egg, larva (nymph), adult.

ORTHOPTERA Grasshoppers & Crickets

Characteristic large hindlegs modified for leaping; usually two sets of wings; large eyes; antennae sometimes very long.

- Incomplete metamorphosis: egg, larva (nymph), adult.

HEMIPTERA True Bugs

United by their sucking mouthparts, narrowed into a 'rostrum', which is like a syringe. Often have partially transparent wings.

- Incomplete metamorphosis: egg, larva (nymph), adult.

NEUROPTERA Lacewings

Delicate, slim insects with two pairs of well-veined wings held roof-wise over body. Long, thin antennae; chewing mouthparts.

- Complete metamorphosis: egg, larva, pupa, adult.

MEGALOPTERA Alderflies

Solid insects with two pairs of heavily veined wings held tent-like over body. Long antennae.

- Complete metamorphosis: egg, larva, pupa, adult.

COLEOPTERA Beetles

Hardened forewings, known as elytra, which usually cover whole back, including hindwings, if they have them; biting jaws (sometimes in larva, too).

- Complete metamorphosis: egg, larva, pupa, adult.

DIPTERA True Flies

Characterized by having just two proper wings, which give them exceptional manoeuvrability; tiny hindwings modified into rod-like halteres, which are sensory organs that give flight information. Eyes large and mouthparts adapted for piercing and sucking.

- Complete metamorphosis: egg, larva, pupa, adult.

MECOPTERA Scorpionflies

Long, simple rostrum (beak) and four membranous wings. Male of most species has sexual organ at tip of abdomen that curves over, making it resemble a scorpion. Primitive, and probably pollinated trees in the Jurassic era (from about 174 million years ago).

- Complete metamorphosis: egg, larva, pupa, adult.

TRICHOPTERA Caddisflies

Two pairs of moth-like but hairy wings; long antennae; adults lack proboscis. Notable for having aquatic larvae, many of which make protective cases.

- Complete metamorphosis: egg, larva, pupa, adult.

LEPIDOPTERA Butterflies & Moths

Celebrated for their two pairs of wings covered in scales, which can be dazzlingly colourful; proboscis for mouthparts, which can be coiled. Butterflies are really just day-flying moths with clubbed antennae.

• Complete metamorphosis: egg, larva, pupa, adult.

HYMENOPTERA Ants, Bees, Wasps & Allies

Two pairs of membranous wings, the forewings larger than the hindwings; mouthparts for biting and chewing; large eyes. Females often have an abdominal tube for egg-laying, the ovipositor, which may be used to sting. Bees are plumper and hairier than wasps.

• Complete metamorphosis: egg, larva, pupa, adult.

GLOSSARY

abdomen Third part of an insect's body, behind thorax.
antenna (pl. **antennae**) Appendage on head, sometimes referred to as 'feeler'. Antennae have several sensory uses, such as feeling and smelling.
aquatic Living in water – many insects' larval stages are fully aquatic, becoming terrestrial when they transition to adult stage.
arista (pl. **aristae**) Bristle on end of a fly's antenna. Can be short and thin, or huge and feathery.
bug Word often used generically to describe all insects, but technically applies to the order Hemiptera ('true' bugs).
cerci Abdominal appendages, such as the pincers on an earwig.
complete metamorphosis Complete change from a larva into an adult by means of a pupa. Larva and adult bear very little resemblance to each other; a larva completely rebuilds itself during the pupal phase. Moths, butterflies, bees, wasps, beetles and flies go through complete metamorphosis.
compound eye Complex eye of an insect, made up of differing quantities of photoreceptive cells, from a few in a lacewing larva eye, to a great many in an adult wasp.
corbicula (pl. **corbiculae**) Also known as 'pollen baskets'. These are the smooth, shiny areas on the hind tibiae where some bees (such as honey bees and bumblebees) attach pollen to carry it back to the hive or nest. They resemble bicycle panniers.
egg First stage in the life cycle of most insects.
elytra Hard casings that cover and protect the wings and soft bodies of beetles. Technically the forewings, although they are not used to fly.
exoskeleton Insects do not have an internal skeleton. Their skeleton is the hardened casing – known as the exoskeleton. It protects the soft inner body and gives an insect its structure. It is made of a substance called chitin.
exuvia Dried, discarded skin shed by insects that moult.
femora First visible section of an insect's leg, closest to body.
frons Part of an insect's face between the eyes.
haltere Small, club-shaped appendage on a fly's thorax. Only found in flies, and helps with balancing and turning in flight.
head First section, and front end, of an insect.
incomplete metamorphosis An insect goes through a series of similar stages from hatching to adult. In incomplete metamorphosis there is no pupal phase; instead it moults through several nymph (instar) phases. Bugs, dragonflies, grasshoppers and bush-crickets go through incomplete metamorphosis.
instar Nymph stage of any insect that goes through incomplete metamorphosis – for example, a shieldbug can have several instars before it moults into its adult form.
larva (pl. **larvae**) Second stage of an insect (that goes through complete metamorphosis), between egg and pupa. The longest phase of the life cycle of many insects. Also often referred to as a 'grub'.
mandibles Curved, pointy appendages around mouth. Used for a variety of purposes, such as catching prey, eating, nest building and fighting.
nymph Stage between egg and adult of insects that

undergo incomplete metamorphosis.
ocelli Simple 'eyes' on top of an insect's head. They do not have lenses and cannot see like compound eyes; instead they sense changes in light intensity and direction. Usually there are three, sometimes fewer; plant bugs (Miridae) do not have any.
ovipositor Egg-laying apparatus of female insects. In some, such as wasps and bees, the ovipositor has evolved to also function as a stinging device.
pile Dense hair on an insect's body.
pollen brush Pollen-collecting apparatus of solitary bees, consisting of very dense tuft of hairs, either on hind tibia, or underneath abdomen.
proboscis Long, retractable feeding tube that extends from mouthparts of insects such as bees, wasps, butterflies and moths.
pronotom Area of exoskeleton that covers thorax.
punctate Refers to pattern on exoskeleton of small, dense, round indentations that gives an appearance of hammered metal.
pupa Third stage of life cycle of insect that undergoes complete metamorphosis. Sometimes referred to as a cocoon. The larva builds a protective, sealed case around itself. In the pupa, the larva essentially liquifies itself, completely rearranging its DNA and body structure to emerge in an utterly different physical state.
rostrum Long, slender, pointed mouthpart of a true bug, such as a shieldbug or damsel bug, used for piercing prey. Does not retract like a bee's tongue; instead it is held underneath the body. In weevils and scorpionflies the rostrum looks like a long nose.
scutellum Section of thorax, just above abdomen. Can be distinctive and aid in identification. Quite large in some bugs, particularly shieldbugs.
sexually dimorphic Refers to species in which male and female look distinctly different.
sp. (pl. **spp.**) Abbreviation for species.
sternites Visible sections underneath an insect's body; underside of tergites.
stigma Thickened section on outside edge of veined wings of insects such as bees, wasps, sawflies, dragonflies and true flies. It can be of different colours, which can help in identification.
tarsus (pl. **tarsi**) End section of leg, furthest away from body; usually segmented.
tergites Segments on top of abdomen.
terrestrial Living on land.
thorax Centre segment of an insect's body (between head and abdomen), to which legs are attached.
tibia (pl. **tibiae**) Section of leg between femora and tarsus.
tympanum Hearing organ of grasshoppers and crickets. Consists of thin membrane located in foreleg joint between femur and tibia, as in the Speckled Bush-cricket.

NOTES ON SPECIES DESCRIPTIONS

The months highlighted in the calendar at the top of the page refer to adult stages only.

This section gives an introduction to the species.

The life cycle of the species. Relevant notes are included, but aren't exhaustive. Occasionally the details even for common insects haven't been discovered.

Fascinating facts.

The area covered includes Britain, Ireland, northern France, Belgium, the Netherlands, Luxembourg, northern Germany, northern Poland, the Baltic States, Scandinavia and Finland. Iceland is not included.

Photos at the top of each page give an indication of where in the garden you can find the insect: pond, herbaceous border, trees, etc. These photos may also help you adapt your garden to attract certain insects.

JAN FEB MAR APR **MAY JUN JUL AUG SEP** OCT NOV DEC

MAYFLIES

Ephemeroptera spp.

WHERE TO FIND

Mayflies are very common in wetlands throughout the region.

For a mayfly, every dance is the last dance. In the spring and summer these delicate insects gather in mating flights over the water. After a brief flurry it is all over – as the females lay eggs on the surface, the males return to land, and all quickly die. Mayflies are moth-like insects, but have translucent wings, of which one pair is vastly larger than the other; they hold them together over the back at rest. They also lack a proboscis and, if all else fails, you can identify them by the beautiful three-pronged tail, also present in the nymphs. The younger stages are entirely aquatic and often last a year or more, so their lives are not as short as people think. Mayflies are also found throughout the warmer months.

EGG (on water's surface) > **LARVA** (nymph) moults many times, the last stage being called a 'sub-imago' > **ADULT.** Larva is aquatic, living on vegetation and organic matter. Adult does not feed.

Mayflies are famous for living only one day in the adult stage, but it is often less than that – just a few hours. Mayflies are the only insects that moult in a winged stage (later stage larva or 'sub-imago', to adult). In May, mayflies are so abundant that fish in waterbodies enter a feeding frenzy and become reckless and easy for anglers to catch.

DRAKE MACKEREL MAYFLY
EPHEMERA VULGATA

FACT FILE

ORDER Ephemeroptera BODY LENGTH 25mm SIMILAR SPECIES Look out for similar, weak-flying insects such as caddisflies (p. 100) and Alderflies (p. 45).

JAN FEB MAR APR **MAY JUN JUL AUG SEP** OCT NOV DEC

BANDED DEMOISELLE

Calopteryx splendens

An unmistakable 'dancing' damselfly, this species flies like a cross between a helicopter and a fairy, exuding a remarkable fluttering grace on four wings. The male's body is a metallic dark blue-green, and it has very obvious dark bands near the wing-tips, which look somewhat like fingerprints. The female is a glittering bronze-green and lacks the bands, looking like a different species. The insects settle in characteristic damselfly pose with wings closed over the back.

EGG (on vegetation underwater) > **LARVA** (underwater, 1–2 years) > **ADULT**. Overwinters as a larva. Both adults and larvae are predatory on small insects.

Females mostly go underwater to lay their eggs and their breathing while submerged is assisted by a trapped layer of air between the wings.

WHERE TO FIND

Common in lowlands throughout the region, except for northern Scotland and Norway; only found in south of Sweden. Mainly a wetland species but wanders into gardens.

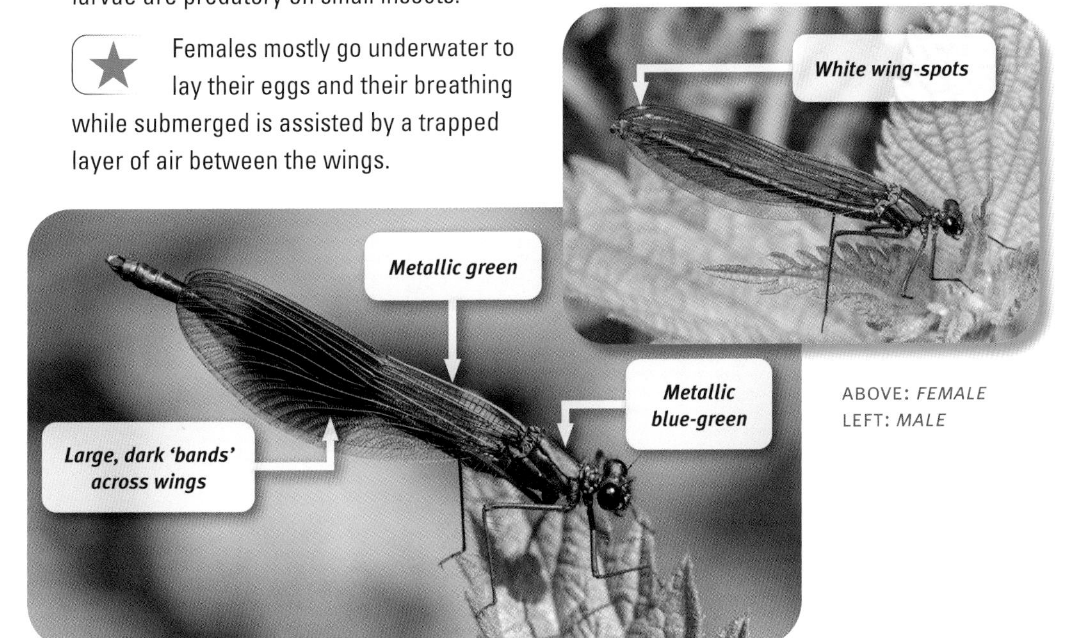

ABOVE: *FEMALE*
LEFT: *MALE*

FACT FILE

ORDER Odonata FAMILY Calopterygidae BODY LENGTH 45–48mm SIMILAR SPECIES Appreciably larger than other damselflies. Grace in flight quite different to rapid forwards darts and sweeps of typical dragonflies. The only other damselfly species with coloured wings is the Beautiful Demoiselle *C. virgo*, which looks quite similar, but its wings are entirely dark and metallic.

JAN	FEB	MAR	**APR**	**MAY**	**JUN**	**JUL**	**AUG**	SEP	OCT	NOV	DEC

LARGE RED DAMSELFLY

Pyrrhosoma nymphula

WHERE TO FIND

Common throughout the region, except in montane and far northern Scandinavia. Visits gardens even in the absence of ponds.

This unmistakable sylph looks like a flying red twig, and on landing resembles a perched red twig. In damselfly fashion, at rest the wings are held closed or nearly closed over the body (not completely open, as in a dragonfly); in flight, it is slow and impossibly delicate. No other common damselfly is this red-brick colour, and it is also the earliest to emerge in spring. The female has yellow bands between the abdominal segments.

EGG (into submerged vegetation, batches of 500) > **LARVA** (aquatic, among weeds, usually for two years) > **ADULT**. Overwinters as a larva. Adults and larvae are predatory.

Only about 0.2 per cent of larvae survive to metamorphose to adults.

ABOVE: *FEMALE*
RIGHT: *MALE*

FACT FILE

ORDER Odonata FAMILY Coenagrionidae BODY LENGTH 33–36mm SIMILAR SPECIES Of similar colour, the male Common Darter (p. 20) is much larger and at rest holds its wings open at right angles to its body.

JAN | FEB | MAR | APR | **MAY** | **JUN** | **JUL** | **AUG** | SEP | OCT | NOV | DEC

AZURE DAMSELFLY

Coenagrion puella

The ethereal powder-blue of this stick-like damselfly (and its near relatives) is unique among common garden insects. It only takes the smallest pond or wet spot to attract it. It frequently perches on stems of waterside plants, wings closed over its back in typical damselfly fashion. Look for the 'U'-shaped (cup-shaped) pattern on the male's second abdominal segment. The female is almost all dark.

EGG (laid into vegetation) > **LARVA** (aquatic, usually for one year in south, two years elsewhere) > **ADULT**.

When mating, damselflies and dragonflies assume a 'cartwheel' position, with the male grabbing the female and holding her tightly by the scruff of the neck with claspers at the tip of his abdomen. If the female accepts him, she will curl the tip of her abdomen to meet his sexual organs under the base of the abdomen. Damselflies can fly when mating. They often remain locked for a while, giving the male time to eject any previous sperm from a rival.

ADULT MALE

WHERE TO FIND

Very common everywhere except the very south of Norway, Sweden and Finland, but there are several similar species that almost reach the Arctic Circle.

FACT FILE

ORDER **Odonata** FAMILY **Coenagrionidae** BODY LENGTH **33–35mm** SIMILAR SPECIES **Common Blue Damselfly *Enallagma cyathigerum* is equally common, but the second abdominal segment has a mushroom-like mark, and the female has blue rings between dark segments. The Blue-tailed Damselfly *Ischnura elegans* has a dark abdomen with a blue segment just before the tail-tip; the female is whitish.**

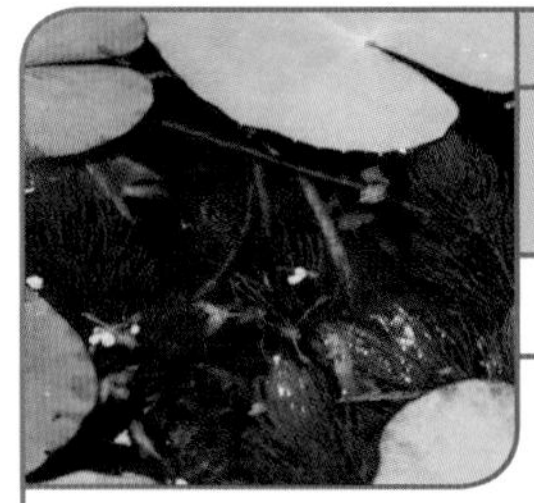

JAN	FEB	MAR	APR	MAY	**JUN**	**JUL**	**AUG**	SEP	OCT	NOV	DEC

EMPEROR DRAGONFLY

Anax imperator

WHERE TO FIND

Common throughout the region, except in the north of Britain and Ireland. Absent from Scandinavia, but similar species thrive up to beyond the Arctic Circle.

This large, powerful, finger-sized, swift-flying dragonfly with attitude is usually seen patrolling low over water, but does visit gardens with modest-sized ponds. It is big and fast enough to be intimidating if it comes too close. The male's abdomen is blue, the colour of a hot summer sky, the thorax emerald-green. The female is largely green all over. Both have a black line down the abdomen. The species is a predator of flying insects, using its long legs to grab them and its formidable jaws to crush them. It will bite people if handled.

EGG (laid on surface of aquatic plants, hatches in three weeks) > **LARVA** (nymph; variable number of stages, submerged in water) > **ADULT** (lives about a month). The later stage nymphs are predators of tadpoles and even small fish.

Unusually for a dragonfly, the Emperor often feeds actively at dusk. Dragonfly larvae shoot water out of their rear ends to propel themselves forwards.

FACT FILE

ORDER Odonata FAMILY Aeshnidae BODY LENGTH 66–84mm SIMILAR SPECIES There are several species of related dragonfly, called hawkers, which might visit gardens. All are smaller than the Emperor. The Common Hawker *Aeshna juncea* has a more spotted abdomen; it reaches the Arctic.

JAN | FEB | MAR | APR | **MAY** | **JUN** | **JUL** | **AUG** | SEP | OCT | NOV | DEC

BROAD-BODIED CHASER

Libellula depressa

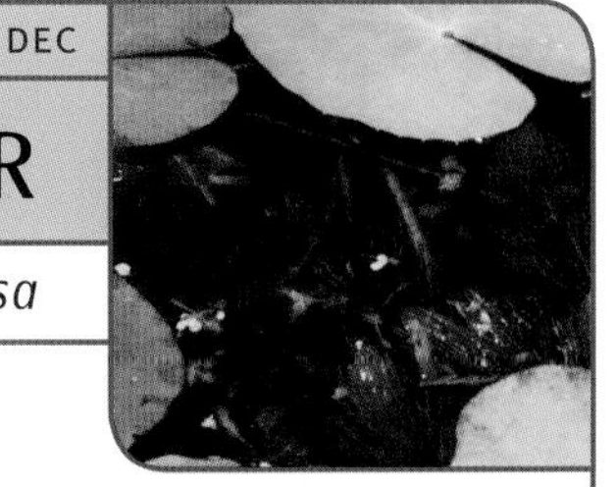

This is a medium-sized dragonfly with an oddly swollen, flattened abdomen, hence its name. It turns up even at the newest garden ponds, and divides its time between low, rapid flights and long periods perched on vegetation, lazing in the sun. The male has a powder-blue abdomen and the female's is yellow and olive-brown, but both have yellow spots on the edges. The bases of the wings are brown.

EGG (laid in flight dipping into water) > **LARVA** (nymph, submerged in mud in water for 1–3 years) > **ADULT**. The later-stage nymphs, as well as the adults, are predators.

If you find identifying dragonflies difficult, spare a thought for the male Broad-bodied Chaser. He sometimes has trouble recognizing females of his own species and tries to mate with the wrong one.

WHERE TO FIND

Common everywhere except Ireland, Scotland and most of Scandinavia. It does occur in southern Sweden and Finland.

ABOVE: *MALE*
LEFT: *FEMALE*

FACT FILE

ORDER Odonata FAMILY Libellulidae BODY LENGTH 39–48mm SIMILAR SPECIES Much smaller than the Emperor Dragonfly (opposite). The Black-tailed Skimmer *Orthetrum cancellatum* has a thinner abdomen with a more extended black tip.

JAN	FEB	MAR	APR	MAY	JUN	**JUL**	**AUG**	**SEP**	**OCT**	NOV	DEC

COMMON DARTER

Sympetrum striolatum

WHERE TO FIND

Common almost everywhere, but more localized in Scandinavia and the eastern Baltic.

This small, intensely orange-red dragonfly, is neat and perfectly proportioned. It often rests for a long time in one spot in the sun, even on the ground. It wanders, and will visit gardens even without a pond, and is characteristic of late summer and autumn. Only the male is reddish; the female and young males are yellow, but still neat and delicate. It has a habit of suddenly darting out from its perch to pursue its prey.

EGG (laid in flight, dipping abdomen tip into water) > **LARVA** (nymphs in mud and aquatic vegetation) > **ADULT**. Overwinters as a predatory larva.

In late summer, immigrants are often blown over on southerly or easterly flows. Some spend the whole journey locked together in a mating embrace.

RIGHT: *FEMALE*
BELOW: *MALE*

FACT FILE

ORDER **Odonata** FAMILY **Libellulidae** BODY LENGTH **33–44mm** SIMILAR SPECIES **Distinguishable from the Large Red Damselfly (p. 16) by its thick abdomen and the way its wings are held out in typical dragonfly style. There are also several similar species of darter.**

JAN | FEB | MAR | APR | MAY | JUN | JUL | AUG | SEP | OCT | NOV | DEC

COMMON EARWIG

Forficula auricularia

Many people are repelled by the alarming looking 'forceps' at the tail end of this familiar garden inhabitant, but it is harmless and cannot pinch humans. It is unmistakable for its long, flattish, dark chocolate-brown body with yellowish folded forewings above the thorax – despite appearances, it can fly. It is abundant everywhere, in flowers, on soil and in compost heaps, feeding on both vegetation and small animals such as aphids, and is often found on walls, fitting snugly into crevices.

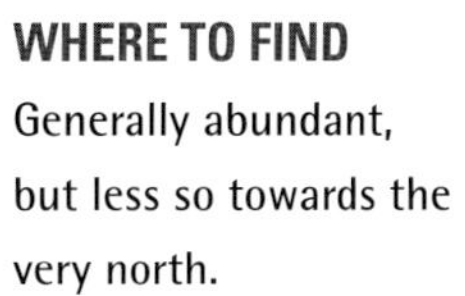

WHERE TO FIND
Generally abundant, but less so towards the very north.

EGG (c. 20–80, laid in underground nest, and looked after by female) > **LARVA** (five stages, early ones also looked after – protected from predators and kept warm) > **ADULT** (lives for one year).

Pity the male. After staying with the female into the winter following egg laying, it is usually summarily evicted from the nest. It dies soon afterwards. Uniquely among British insects, mother earwigs are fastidious about cleanliness, constantly attending to the hygiene of the eggs.

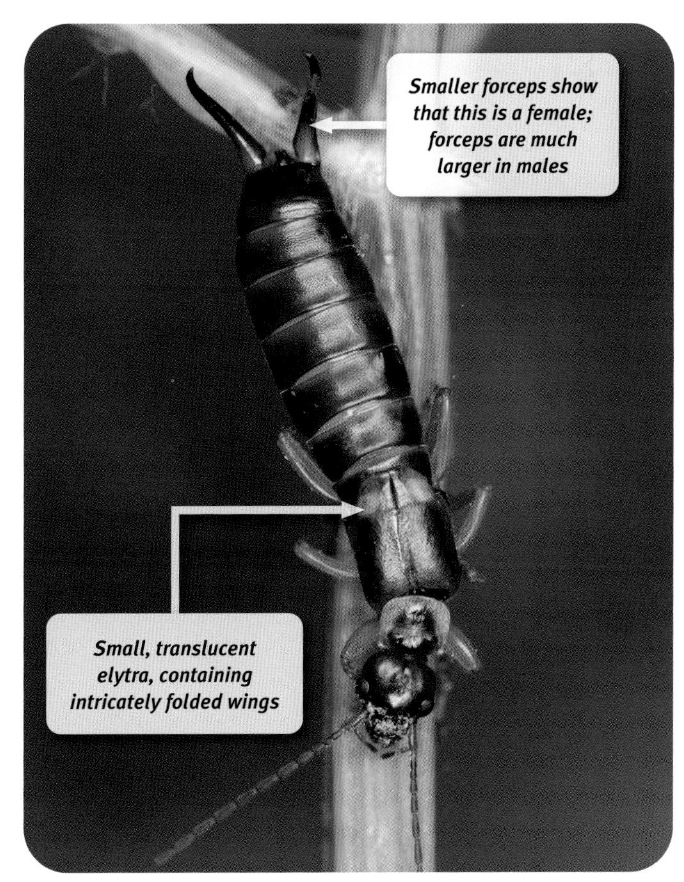

FACT FILE

ORDER Dermaptera FAMILY Forficulidae BODY LENGTH 10–15mm
SIMILAR SPECIES None.

JAN	FEB	MAR	APR	MAY	JUN	**JUL**	**AUG**	**SEP**	**OCT**	**NOV**	DEC

OAK BUSH-CRICKET

Meconema thalassinum

WHERE TO FIND

Common in the southern half of Britain and throughout much of the region north to southern Sweden.

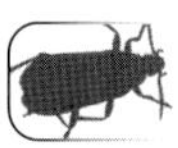

Few people would imagine that this wonderful insect roams their night-time gardens. The almost luminous, plastic-green colouration, the eyelash-thin antennae that are as long as the body, the pointed rear end (ovipositor in female, pincer-like cerci in male), as well as the flexed legs, add up to an impressive sight. This animal is nocturnal, often attracted to window lights (and moth traps). It flies strongly. The male 'sings' by tapping its foot on a leaf, unlike other crickets, which rub their legs/wings together to 'sing'. The song is barely audible to humans.

EGG (laid one by one in bark and other crevices, using female's long ovipositor) > **LARVA** (nymph, five instars) > **ADULT**. The nymphs and adults feed on small insects.

Unlike most other crickets and grasshoppers, it lives in trees and bushes.

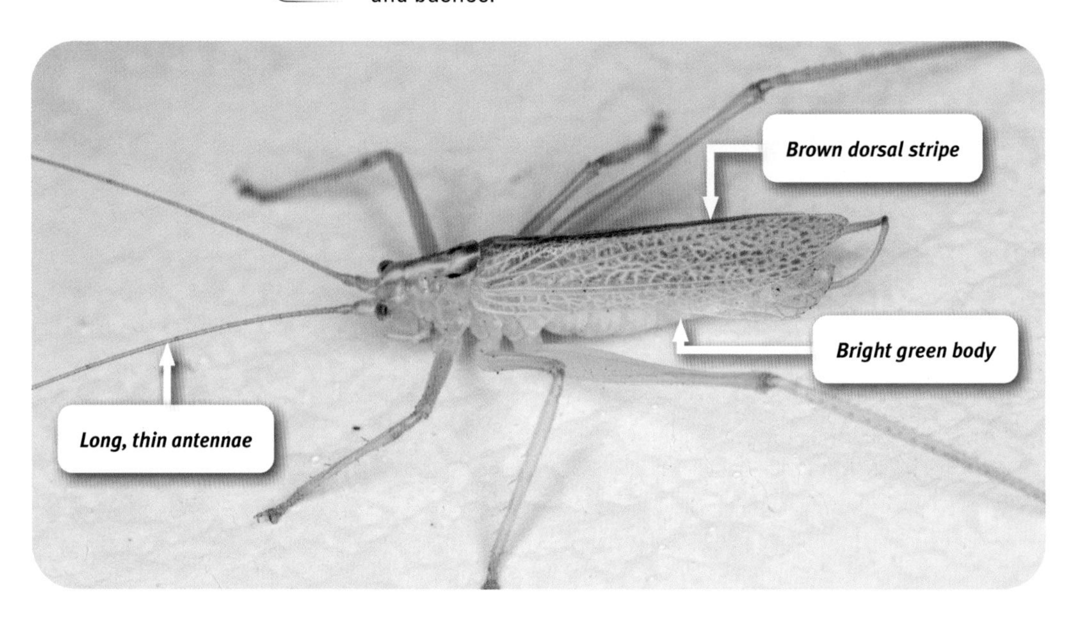

FACT FILE

ORDER Orthoptera FAMILY Tettigoniidae BODY LENGTH 13–17mm SIMILAR SPECIES Larger than a grasshopper and those antennae are a giveaway. There are several other bush-crickets in Britain. The Speckled Bush-cricket (opposite) is wingless.

JAN FEB MAR APR MAY JUN **JUL AUG SEP OCT NOV** DEC

SPECKLED BUSH-CRICKET

Leptophyes punctatissima

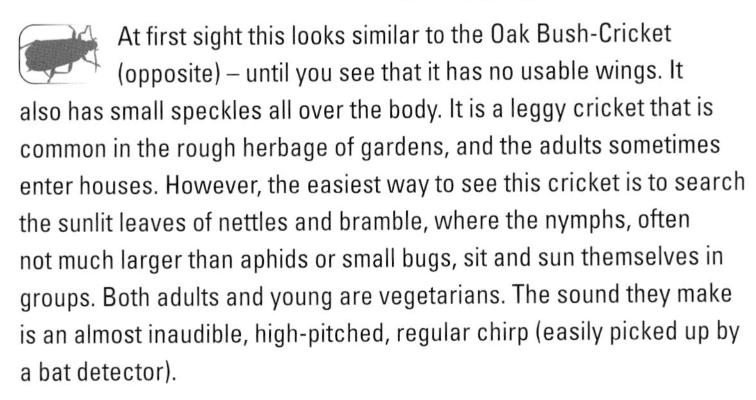

At first sight this looks similar to the Oak Bush-Cricket (opposite) – until you see that it has no usable wings. It also has small speckles all over the body. It is a leggy cricket that is common in the rough herbage of gardens, and the adults sometimes enter houses. However, the easiest way to see this cricket is to search the sunlit leaves of nettles and bramble, where the nymphs, often not much larger than aphids or small bugs, sit and sun themselves in groups. Both adults and young are vegetarians. The sound they make is an almost inaudible, high-pitched, regular chirp (easily picked up by a bat detector).

WHERE TO FIND
Common in southern England and up to the Scottish borders, while elsewhere it is widespread up to southern Sweden. Restricted to the south of Ireland.

EGG (laid one by one in bark and other crevices, using female's ovipositor) > **LARVA** (nymphs, in vegetation) > **ADULT**. The nymphs feed on plants, including nettles and bramble. Overwinters as an egg.

It makes its chirp by rubbing the wings together.

ABOVE: *NYMPH;*
LEFT: *ADULT*

FACT FILE

ORDER Orthoptera FAMILY Tettigoniidae BODY LENGTH 9–18mm SIMILAR SPECIES The Oak Bush-Cricket has wings and a longer ovipositor.

JAN	FEB	MAR	APR	MAY	**JUN**	**JUL**	**AUG**	**SEP**	**OCT**	**NOV**	DEC

COMMON FIELD GRASSHOPPER

Chorthippus brunneus

WHERE TO FIND
Widespread and generally very common.

Grasshoppers are the familiar spring-loaded jumpers of the grass of summer, whose chirping (stridulation) forms a warm-weather soundtrack. They are discouraged by well-mown lawns, but leap at the prospect of a meadow or wild patch in the garden, which they populate by flying in from elsewhere. This is the most likely species to colonize. It can be green or brown, striped or mottled. Males have red tips to the abdomen. Grasshoppers are vegetarian in diet. The sound this species makes is two chirps a second in series of 8–20.

EGG (laid in batches on soil within a pod of hardened froth formed by a secretion from the female) > **LARVA** (nymph, in grass) > **ADULT**.

This grasshopper is particularly abundant on the central reservations of motorways, benefiting from the high levels of nitrogen caused by pollution.

FACT FILE

ORDER Orthoptera FAMILY Acrididae BODY LENGTH 15–25mm SIMILAR SPECIES Shorter antennae than those of crickets. There are many other species of grassshopper in the region.

JAN FEB **MAR APR MAY JUN JUL AUG SEP OCT NOV** DEC

BIRCH SHIELDBUG

Elasmostethus interstinctus

In common with all the shieldbugs, this insect has a thickened plate between the wing-cases called the scutellum, which looks like a Roman shield. Although beetle-like, a bug's mouthparts are a tube (a rostrum) rather than jaws. This is one of several very attractive green shieldbugs that have reddish-brown patterns on the back, with narrow lines that look like a Greek letter. Not surprisingly, this species is most common around birch trees, but it will use other trees and commonly visits large gardens.

WHERE TO FIND
Generally common throughout the region.

EGG (batches laid on leaves) > **LARVA** (nymphs, five different stages) > **ADULT**. Nymphs drink sap. Overwinters as an adult in leaf litter.

Digests sap using bacteria. Females cover their eggs with bacteria so that offspring ingest them while feeding on the egg case.

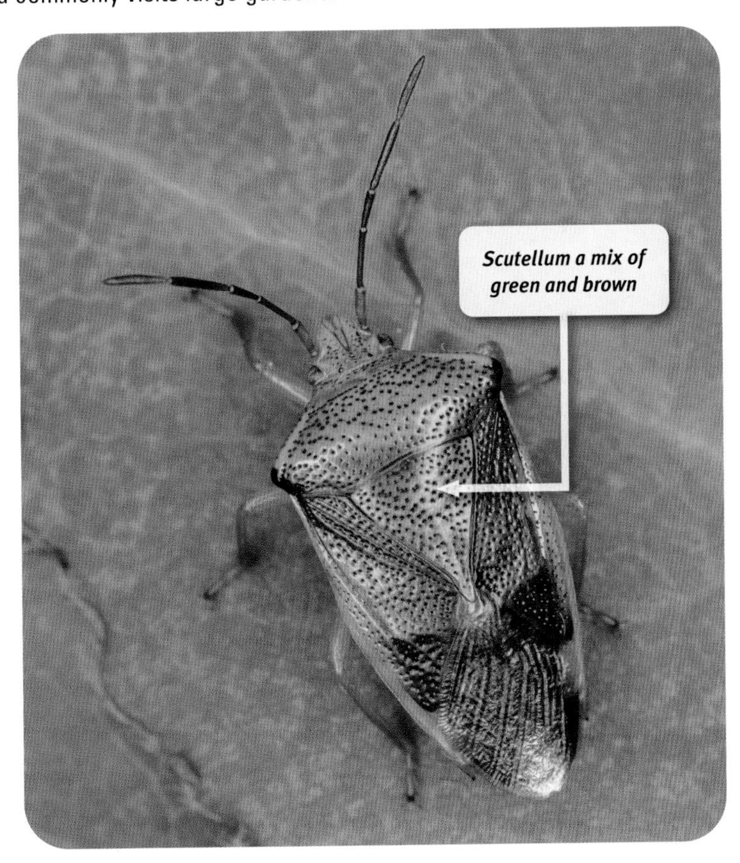

FACT FILE

ORDER Hemiptera FAMILY Acanthosomatidae BODY LENGTH 8–12mm SIMILAR SPECIES The Hawthorn Shieldbug (p. 26) is similar but larger. The scutellum is green in the Hawthorn Shieldbug, green and brown in the Birch Shieldbug.

JAN FEB **MAR APR MAY JUN JUL AUG SEP OCT NOV** DEC

HAWTHORN SHIELDBUG

Acanthosoma haemorrhoidale

WHERE TO FIND

Very common and widespread species occurring wherever hawthorn grows.

It is easy to see why shieldbugs get their name, from the hardened plates covering the body. They are common insects, although very well camouflaged – in the Hawthorn Shieldbug the beautiful brown markings on the back break up the animal's profile. Half the wing is leathery, a feature of the group of insects known as true bugs (Hemiptera), but they fly well. This is an abundant species found not only on hawthorn but on many other plants. The broad brown markings are a useful identification pointer, as are the sideways extensions of the shield (they look like eyes in the photograph).

EGG (batches laid on leaves, usually of hawthorn) > **LARVA** (nymphs, five different stages, eat hawthorn berries) > **ADULT**. Adults eat leaves. Overwinters as an adult.

These insects can release foul-smelling liquids from their bodies as a defence.

FACT FILE

ORDER Hemiptera FAMILY Acanthosomatidae BODY LENGTH 12–16mm SIMILAR SPECIES The Birch Shieldbug (p. 25) is less colourful, with narrow reddish-brown markings.

JAN FEB MAR APR MAY JUN **JUL AUG SEP OCT NOV** DEC

RED-LEGGED SHIELDBUG

Pentatoma rufipes

The 'Roman shield' body (pronotum) of this handsome dark-brown bug, also known as the Forest Shieldbug, has a slight hook on either side, setting it apart from its relatives. It also has a very distinctive orange or cream spot in the middle of the shield (technically at the tip of the scutellum), which makes it straightforward to identify. The amount of red on the legs varies, from completely dull to magnificent bright red. A common bug that drinks the sap from almost any tree in the garden, it will also turn its attention to piercing fruits and the flesh of caterpillars and other insects.

WHERE TO FIND

Very common throughout the region, mainly on oak trees.

EGG (batches of 10–15 laid on leaves) > **LARVA** (nymphs, five stages) > **ADULT**. Nymphs drink sap. Eggs are laid in August and nymphs overwinter.

Shieldbugs make a loud, rasping buzz in flight that distinguishes them from bees and flies.

Red-brown legs

Metallic blue and green markings on pronotum and abdomen.

Orange or cream tip to scutellum

LEFT: *NYMPH*; RIGHT: *ADULT*

FACT FILE

ORDER **Hemiptera** FAMILY **Pentatomidae** BODY LENGTH **12–14mm** SIMILAR SPECIES Several other shieldbugs occur in gardens, but the shape of this one's shield is distinctive.

JAN FEB **MAR APR MAY JUN JUL AUG SEP OCT NOV** DEC

GREEN SHIELDBUG

Palomena prasina

WHERE TO FIND
Common throughout most of the region to the fringes of the Arctic. In Britain, occurs mainly in the southern half of the country.

It is a shieldbug and it is green. The wing membranes are brown and not clear. This is one of the most common larger shieldbugs, abundant on all kinds of herbage, including trees and shrubs.

EGG (batches of 100 laid on leaves) > **LARVA** (nymphs, five stages) > **ADULT**. Nymphs drink sap. Overwinters as an adult.

FACT FILE

ORDER Hemiptera FAMILY Pentatomidae BODY LENGTH 11–14mm SIMILAR SPECIES Other green shieldbugs have brown or red patches, or clear wing membranes.

JAN FEB MAR APR **MAY JUN JUL AUG SEP OCT NOV** DEC

DOCK BUG

Coreus marginatus

WHERE TO FIND
Common and widespread in most of the region, but only the southern parts of Britain, Ireland and Scandinavia.

The Dock Bug belongs to a group known as the leatherbugs, due to their leathery appearance. They look a little like shrivelled shieldbugs that have been flattened, making their abdomen rounded. They are common on docks and sorrels.

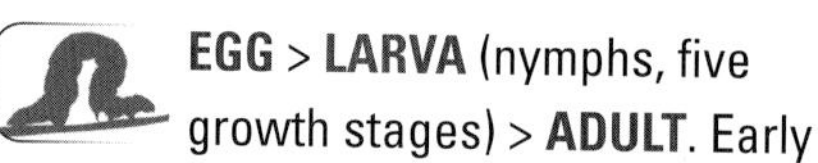

EGG > **LARVA** (nymphs, five growth stages) > **ADULT**. Early nymphs feed on leaves and stems, while late nymphs and adults feed on seeds. Overwinters as an adult.

FACT FILE

ORDER Hemiptera FAMILY Coreidae BODY LENGTH 13–15mm SIMILAR SPECIES Shieldbugs are generally broader beamed and even, without the bulbous abdomen.

JAN | FEB | MAR | APR | MAY | JUN | JUL | AUG | SEP | OCT | NOV | DEC

CINNAMON BUG

Corizus hyoscyami

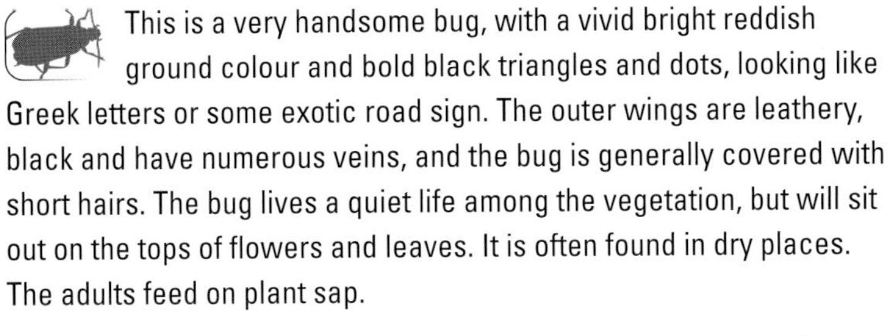

This is a very handsome bug, with a vivid bright reddish ground colour and bold black triangles and dots, looking like Greek letters or some exotic road sign. The outer wings are leathery, black and have numerous veins, and the bug is generally covered with short hairs. The bug lives a quiet life among the vegetation, but will sit out on the tops of flowers and leaves. It is often found in dry places. The adults feed on plant sap.

EGG (on plants) > **LARVA** (nymph, feeds on vegetation) > **ADULT**. Can overwinter as an adult or an egg.

This bug is very unusual for 'singing' during courtship, rubbing its legs against its back. It is much less smelly than other bugs when disturbed.

WHERE TO FIND

Widespread and common north to southern Scandinavia. In Britain, found in southern England as far north as Yorkshire; expanding.

FACT FILE

ORDER Hemiptera FAMILY Rhopalidae BODY LENGTH 9mm SIMILAR SPECIES Several ground bugs (Lygaeidae) are similar, as well as the Red-and-black Froghopper (p. 39) and beetles.

JAN | FEB | MAR | APR | MAY | JUN | JUL | AUG | SEP | OCT | NOV | DEC

BIRCH CATKIN BUG

Kleidocerys resedae

WHERE TO FIND
Common throughout most of the region.

This is one of those insects that you might never notice over a lifetime, part of the small change of the garden insect fauna. It lives much of its life on birch leaves and catkins, and sometimes alder. It uses its mouthparts – which, like those of other true bugs (order Hemiptera), have evolved into a beak (rostrum) that punctures surfaces like a hypodermic needle – to pierce seeds and leaves. This is a small but pleasingly rusty-brown species with largely transparent wings.

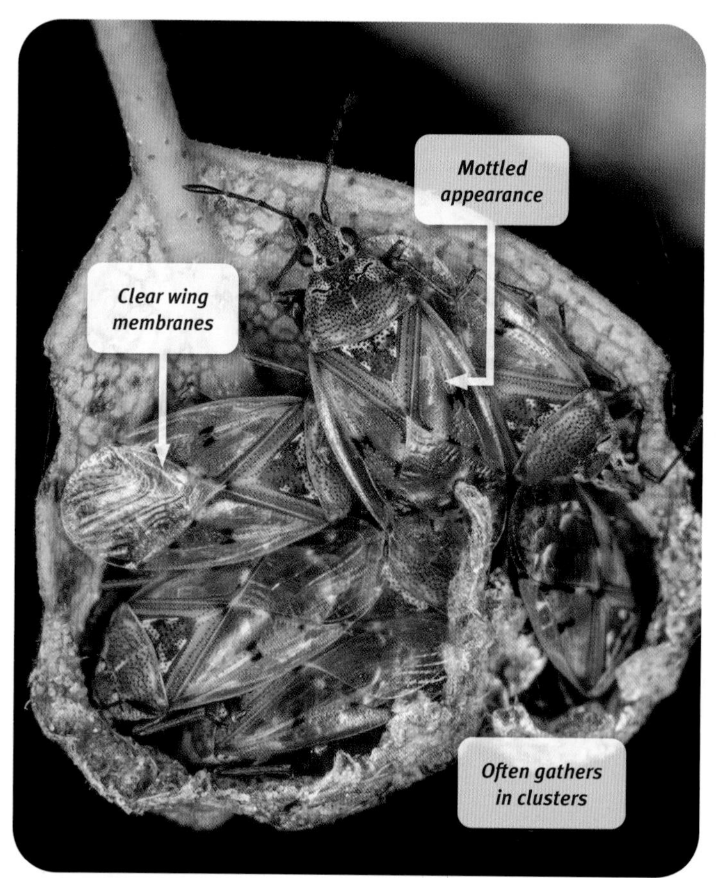

CLUSTER OF BIRCH CATKIN BUGS

EGG (batches on leaves of birch) > **LARVA** (nymph, on birch seeds) > **ADULT**. Nymphs and adults eat seeds from birch catkins, and probably other plant juices. Overwinters as an adult.

In November, drops down from the trees to the ground to hibernate; it can survive even if it lands on snow. Often hangs out in large 'gangs'.

FACT FILE

ORDER Hemiptera FAMILY Lygaeidae BODY LENGTH 4.5–5.5mm SIMILAR SPECIES There are many species of bug in the garden. This one looks like a miniature shieldbug.

JAN | FEB | MAR | APR | MAY | JUN | JUL | AUG | SEP | OCT | NOV | DEC

LACEBUGS

Tingidae spp.

Sometimes small is beautiful, and the intricate markings on this 4mm-long lacebug are worthy of comparison to a mosaic from the ancient world – or indeed, lacework. They could also be called tortoise bugs, with their heavy armour covering the body – albeit it would be a very flat tortoise. These beautiful midgets are suckers of plant juices and can occasionally be a nuisance, but the species illustrated is best looked for on thistles – in fact, only thistles, because lacebugs are host specific.

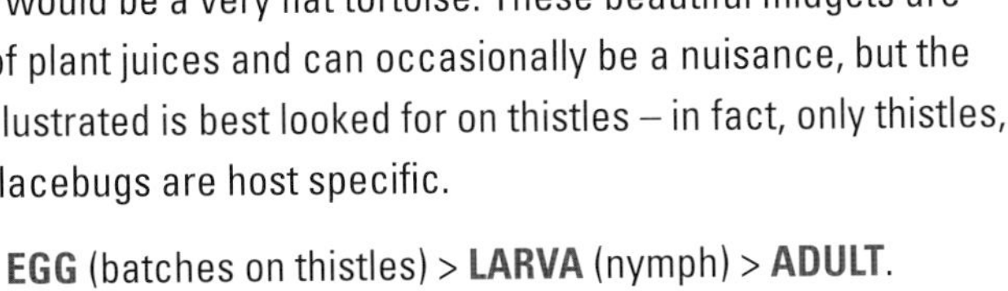

EGG (batches on thistles) > **LARVA** (nymph) > **ADULT**. Nymphs and adults feed on juices of thistles. Overwinters as an adult.

Lacebugs often complete their entire life cycle on one part of one plant.

WHERE TO FIND

Lacebugs are common throughout the region short of the far north, but easily overlooked.

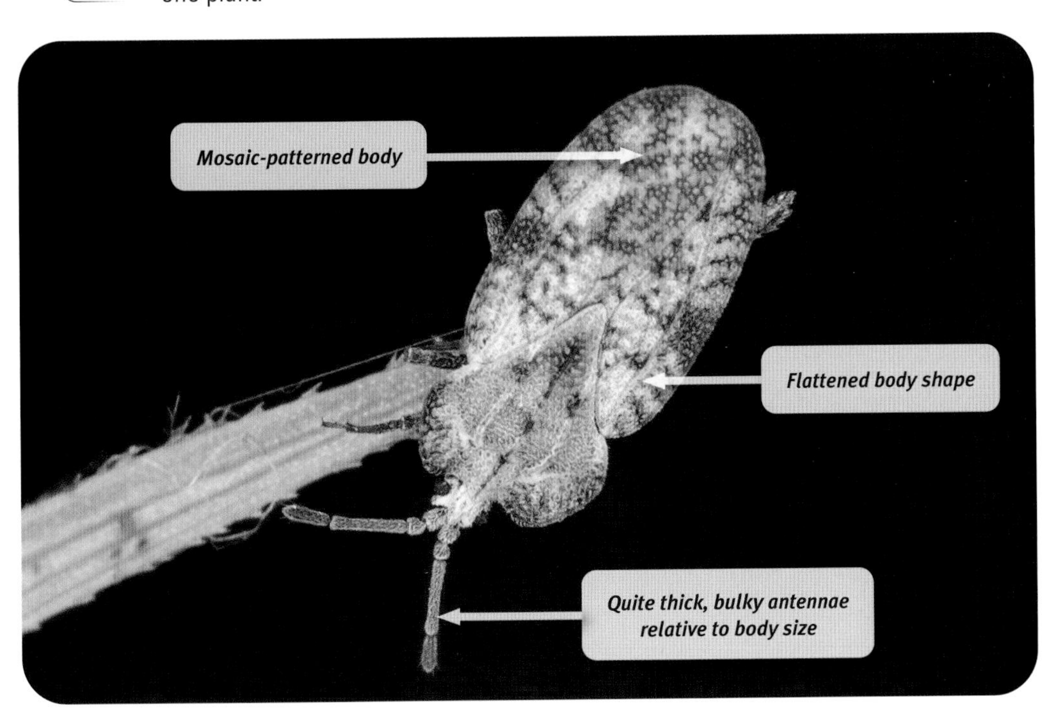

TINGIS AMPLIATA

FACT FILE

ORDER **Hemiptera** FAMILY **Tingidae** BODY LENGTH **3–4mm** SIMILAR SPECIES **Distinctive.**

JAN | FEB | MAR | APR | MAY | JUN | **JUL** | **AUG** | **SEP** | **OCT** | NOV | DEC

TREE DAMSEL BUG

Himacerus apterus

WHERE TO FIND

Common on the Continent, including in the Baltic States, but not found in Scandinavia or Ireland. In the UK, mainly found in southern England.

This is a foe to strike terror into small insects' hearts. Despite its delicate name, it is an indiscriminate slaughterer of aphids, mites, small spiders and anything of diminutive size, walking up on them, piercing their flesh with its sharp mouthparts and sucking them dry. This reddish, spidery species lives in deciduous trees, hunting at all levels; the best way to see one is to beat a branch (p. 7). Interestingly for a tree-dwelling insect species, it cannot fly, because its wings are much reduced in size.

EGG (on plant stems) > **LARVA** (nymph) > **ADULT**. Nymphs and adults highly predatory. Overwinters as an egg.

★ Not at all choosy in its diet, and will eat anything alive smaller than itself.

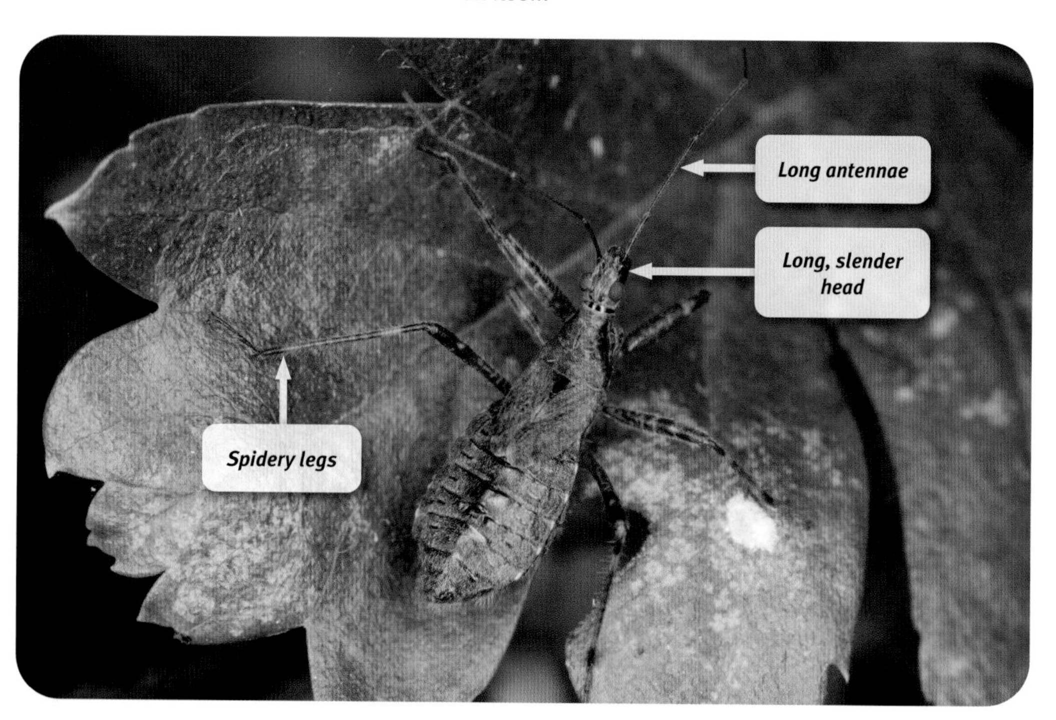

FACT FILE

ORDER **Hemiptera** FAMILY **Nabidae** BODY LENGTH **8–12mm** SIMILAR SPECIES **Other damsel bugs are of similar shape but live on the ground. It looks rather like a giant aphid.**

JAN | FEB | MAR | APR | MAY | JUN | JUL | AUG | SEP | OCT | NOV | DEC

COMMON FLOWER BUG

Anthocoris nemorum

This very small, attractively marked bug packs a punch, being capable of biting you. It looks as though it has a tiny straw permanently attached to its mouth, but those mouthparts are like a miniature syringe and can pierce human skin to drink blood. Its day job is to suck the juices out of aphids (greenflies), mites and other still tinier insects. This is a truly abundant bug found on every kind of vegetation, from ground level to the tops of trees, although it prefers the former. Its black head and upper thorax are shiny, as if polished, and it has shiny forewings. The overall pattern is spotted, often with an ornate orange or brownish pattern. It has an hourglass pattern at the end of the wings.

WHERE TO FIND
Widespread and common throughout the region.

EGG (inside plant leaves) > **LARVA** (nymph) > **ADULT**. Overwinters as an adult, behind bark or in leaf litter.

It is sometimes used as a pest-control agent for aphids.

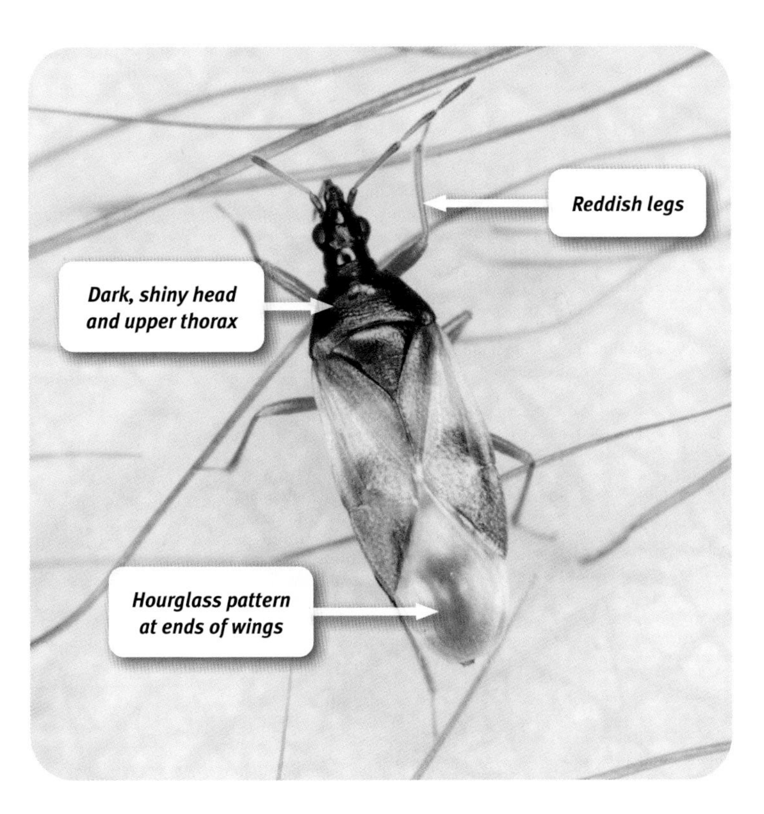

FACT FILE

ORDER **Hemiptera** FAMILY **Anthocoridae** BODY LENGTH **3–4mm** SIMILAR SPECIES **Other small truc bugs (Hemiptera).**

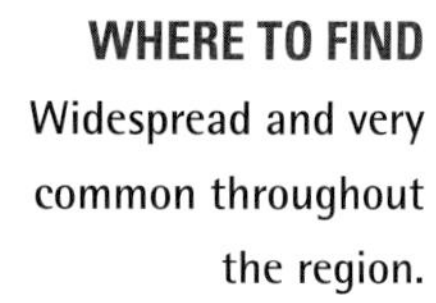

JAN	FEB	MAR	APR	MAY	**JUN**	**JUL**	**AUG**	**SEP**	**OCT**	NOV	DEC

COMMON GREEN CAPSID

Lygocoris pabulinus

WHERE TO FIND
Widespread and very common throughout the region.

Looking very like a shieldbug, although one that has clearly been on a crash diet, this insect is much slimmer and also has considerably longer legs and antennae. The green back (the pronotum) seems also to have been recently washed and waxed. Note the peculiar white eyes. These bugs are abundant and unfussy, content with all kinds of plant juices. Unfortunately, these can include garden and commercial fruit juices.

EGG (200 per female, on woody plants such as fruit trees) > **LARVA** (nymph, moves to herbaceous plants such as nettles) > **ADULT**. Overwinters as an adult, often in leaf litter.

The males court by vibrating their abdomen. They will do this when they come across dead females, or even just female leg body parts.

FACT FILE

ORDER **Hemiptera** FAMILY **Miridae** BODY LENGTH **5–6.5mm** SIMILAR SPECIES **See shieldbugs; there are other green bugs, too.**

JAN | FEB | MAR | APR | MAY | **JUN** | **JUL** | **AUG** | **SEP** | OCT | NOV | DEC

MEADOW PLANT BUG

Leptopterna dolabrata

This is a handsome and abundant insect that you could easily find in the garden simply by sweeping a net over long grass in the summer. It feeds on the developing seeds of grasses of many kinds, and if it reaches commercial crops, such as wheat, it can be a problem. It is unusual among bugs for the fact that the male and female look different, the males being more boldly coloured. Oddly, the males almost always have fully developed wings, while the females' wings may be long or short.

WHERE TO FIND
Widespread and common in most of the region.

EGG (laid at bases of grass stems, female making holes for eggs) > **LARVA** (nymph, hatches in spring and moults five times) > **ADULT**. Overwinters as an egg.

Short-winged females produce many more eggs than their longer winged counterparts.

FEMALE

FACT FILE

ORDER Hemiptera FAMILY Miridae BODY LENGTH 8–10mm SIMILAR SPECIES There are many species of plant bug with very similar body shapes. The closely related Red Meadow Plant Bug *L. ferrugata* is redder and less boldly marked, and occurs in drier areas.

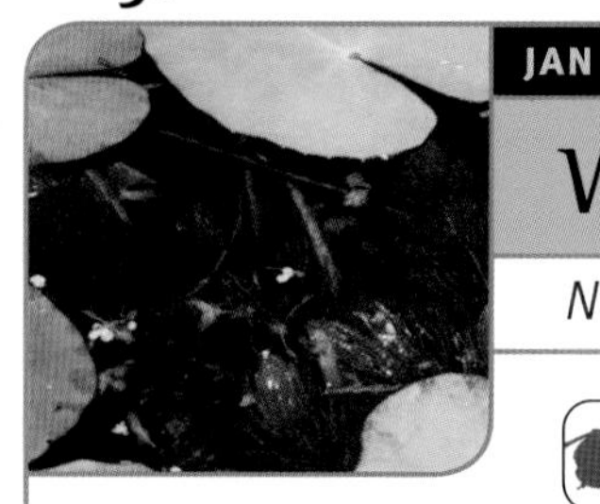

JAN | FEB | MAR | APR | MAY | JUN | JUL | AUG | SEP | OCT | NOV | DEC

WATER SCORPION

Nepa cinerea

WHERE TO FIND
Fairly common in garden ponds throughout the region.

This amazing insect often steals the show from newts and frogs on pond-dipping sessions. It looks like a scorpion, with its remarkable long pincers (front legs in this case), which are used to grab prey, and for the equally striking long tail, which is actually a hollow breathing tube. It has a flattened body, and its dark colour serves as camouflage. It lives in the detritus at the bottom or side of a pond, walking furtively, and catches fish, tadpoles and other live prey. Adults sometimes come out on land at night, and remarkably, it is able to fly.

EGG (laid among algae and plants below the water's surface) > **LARVA** (nymph) > **ADULT** Overwinters as an adult. Breeds in April and May.

★ It has a brilliant red abdomen, presumably to startle predators, but it hardly ever opens its wings to show it off.

FACT FILE

ORDER Hemiptera FAMILY Nepidae BODY LENGTH 17–23mm SIMILAR SPECIES There are no similar species.

JAN | FEB | MAR | **APR** | **MAY** | **JUN** | **JUL** | **AUG** | **SEP** | **OCT** | **NOV** | DEC

COMMON PONDSKATER

Gerris lacustris

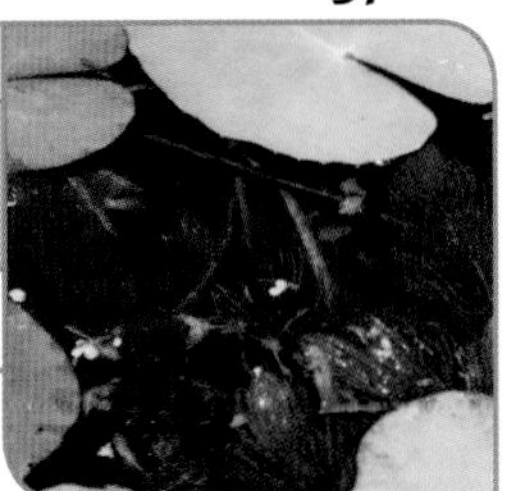

Even the smallest, newest garden pond will attract this, one of our most remarkable and easily identified insects. It rides the surface tension of the water in search of prey, whose misadventure may land it accidentally in water, and whose struggles are akin to shaking a spider's web. Its long central legs allow this bug to skate rapidly over the surface, using a form of breaststroke. The hindlegs act as rudders and the forelegs grab hold of the food and bring it flush to the piercing mouthparts. Some individuals, but not all, are fully winged and can fly strongly, usually at night, easily populating new waterbodies.

WHERE TO FIND

Widespread and common in garden ponds.

EGG (in water) > **LARVA** (nymph, five growth stages over a month) > **ADULT**. Adults overwinter on land and can live for six months.

The pondskater can run across the water's surface at 1.5m per second. The legs have water-resistent hairs. Pondskaters wash themselves by flipping over, 'like self-tossing pancakes'.

FACT FILE

ORDER Hemiptera FAMILY Gerridae BODY LENGTH 8–10mm SIMILAR SPECIES There are other closely related species.

JAN | FEB | MAR | APR | MAY | JUN | JUL | AUG | SEP | OCT | NOV | DEC

COMMON BACKSWIMMER

Notonecta glauca

WHERE TO FIND

Common in garden ponds throughout the region.

The name sums up this remarkable bug: it lives in garden ponds and swims upside down, using its feathery hindlegs to propel it forwards through the water in rowing fashion at considerable speed. It is sometimes called the Water Boatman (of a sunk boat!). It is a serious predator of other animals up to the size of tadpoles and small fish, which it attacks with its needle-like rostrum, inserting it and sucking up body fluids. The large eyes make it look as though it is wearing swimming goggles. Much of its prey is taken near the surface, where land-based insects often find themselves accidentally marooned. The animal is notorious for delivering a painful wound to pond dippers, so needs to be treated with care.

EGG (in plant stems) > **LARVA** (nymph, five growth stages in the water) > **ADULT**. Nymphs are predators. Overwinters as an adult. Nymphs mature into adults in late summer.

Common Backswimmers can distinguish between the different surface waves made by struggling prey, by their own kind and by inanimate objects falling into the water. When flying, they often mistake reflective car windscreens for ponds and try to land on them, only to bounce off.

FACT FILE

ORDER Hemiptera FAMILY Notonectidae BODY LENGTH 14mm SIMILAR SPECIES The Lesser Water Boatman (*Corixa* genus) has a similar swimming style, but is smaller and is also vegetarian.

JAN | FEB | MAR | **APR** | **MAY** | **JUN** | **JUL** | **AUG** | SEP | OCT | NOV | DEC

RED-AND-BLACK FROGHOPPER

Cercopis vulnerata

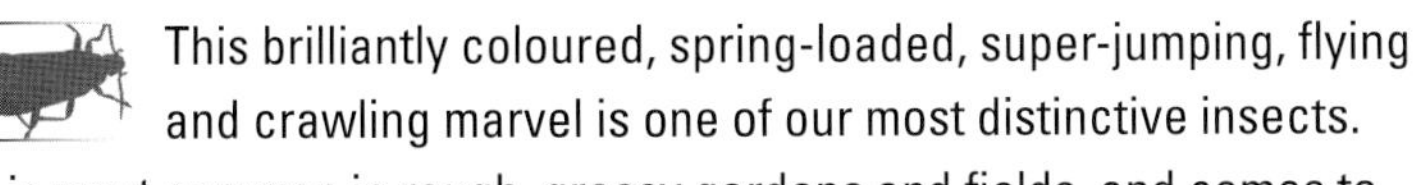

This brilliantly coloured, spring-loaded, super-jumping, flying and crawling marvel is one of our most distinctive insects. It is most common in rough, grassy gardens and fields, and comes to notice as it leaps away from trampling feet. A closer look might reveal it on stems of grasses and low-growing vegetation, where it feeds on sap. Its markings indicate to predators that it is not very tasty.

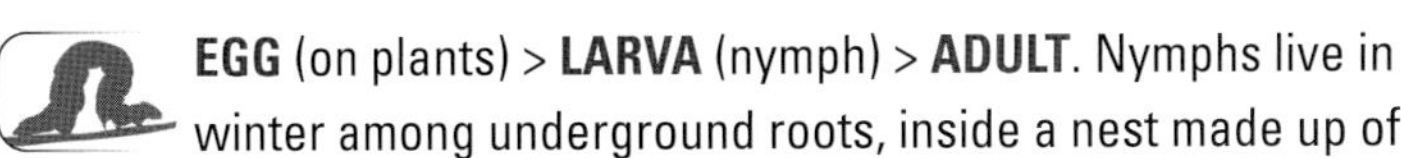

EGG (on plants) > **LARVA** (nymph) > **ADULT**. Nymphs live in winter among underground roots, inside a nest made up of foam. Nymphs feed on root sap.

This froghopper can jump up to 70cm, seven times its own height, and accelerate at 4000m/s^2. It mates in a somewhat tasteful side-to-side posture, and both sexes look as though they could be doing something else at the same time.

WHERE TO FIND

Widespread and common, but absent from Scotland, Ireland and northern Scandinavia.

FACT FILE

ORDER Hemiptera FAMILY Cercopidae BODY LENGTH 9–11mm SIMILAR SPECIES See the differently shaped Cinnamon Bug (p. 29). This froghopper may be mistaken for a ladybird or other colourful beetle, but its distinctive shape and ability to jump distinguishes it.

JAN	FEB	MAR	APR	**MAY**	**JUN**	**JUL**	**AUG**	**SEP**	OCT	NOV	DEC

COMMON FROGHOPPER

Philaenus spumarius

WHERE TO FIND
Common throughout the region.

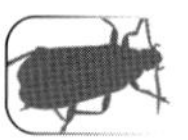

This tiny inhabitant of grass and borders would live entirely under the radar were it not for its remarkable talent – flatulence. It is easy to overlook the insect itself hidden among the leaves; it is a small bug with leathery wings and the ability to hop mighty distances. However, the early stages, the nymphs, create conspicuous foam shelters that look a little like spittle, earning them the country name 'cuckoo spit'. They are formed as the nymphs feed on plant juices (from the main stem 'artery', the xylem), expelling them under pressure from their rear ends to form the froth. Look out for the many different adult colour forms. They can be different shades of black, white and brown.

EGG > **LARVA** (nymph) > **ADULT**. Nymphs live enveloped in a protective splodge of foam, which looks like human saliva (often called cuckoo spit.)

★ The cuckoo-spit froth comes from the rear end, so the frothy shelter is basically made by the insect expelling air into its food.

ABOVE: *CUCKOO-SPIT FROTH*

FACT FILE

ORDER Hemiptera FAMILY Aphrophoridae BODY LENGTH 5–7mm SIMILAR SPECIES Similar to leafhoppers, but the wings have rougher surfaces.

JAN | FEB | MAR | APR | MAY | JUN | **JUL** | **AUG** | **SEP** | **OCT** | NOV | DEC

COMMON GREEN LEAFHOPPER

Cicadella viridis

A leafhopper is one of those insects that lives in grass and leaps away from danger using its spring-loaded legs, although it can also fly readily. It has a compact body, with wings neatly cloaking it. For those familiar with cicadas, it looks like a much-reduced version of a cicada, hence the generic name *Cicadella*. This is a distinctive species in which the female (as here) is fresh green, but the male is bluish-purple. A useful feature of this species is that the head section is both yellow and green; the markings might even recall the head of a budgerigar, the bird. It feeds on plant sap.

EGG (laid in autumn) > **LARVA** (nymph) > **ADULT**. Nymphs feed on sap. Overwinters as an egg.

Leafhoppers exude bronchosomes, unique granules that are super-repellent of water and protect them from their own sticky excreta.

WHERE TO FIND

Abundant in grassy places, especially damp ones, throughout the region.

FACT FILE

ORDER Hemiptera FAMILY Cicadellidae BODY LENGTH 6–8mm SIMILAR SPECIES Froghoppers are similar. There are many different leafhoppers, with different colour variants. Most have rows of spines on the legs (tibiae). This differentiates them from froghoppers, which have only two spines in this area.

JAN | FEB | MAR | APR | MAY | JUN | JUL | AUG | SEP | OCT | NOV | DEC

APHIDS

Aphididae spp.

WHERE TO FIND

Abundant everywhere, although there are fewer in the far north.

Aphids are among the least popular of garden insects, but definitely among the most fascinating. Everyone knows them, often as greenflies or blackflies. They are very small, with a pear-shaped body – a small head and bulbous abdomen. There are unwinged and winged forms, often of the same species, and their wings are unusual for having one main vein along the leading edge, with a few branches. They spend their lives drinking plant sap with their sharp mouthparts, often in large numbers, and it is these multitudes that cause damage to favourite garden plants such as roses. They drink so much that liquid is quickly exuded from the rear end as honeydew, and is frequently a great attractant for ants, which 'farm' the smaller insects. In midsummer many aphids take to the sky and may be wafted across enormous distances.

EGG (often none) > **LARVA** (nymph, may be born live) > **ADULT**. Adults and nymphs feed on sap.

Females often give birth to live nymphs (parthenogenesis), which themselves might already be pregnant. This is known as telescopic development. Winged females tend to develop late in the season and may colonize different species of plant from those used by their female parent. The Yellow Meadow Ant (p. 142) manages large herds of aphids.

LEFT: *ANT FARMING*; RIGHT: *APHID*

FACT FILE

ORDER Hemiptera FAMILY Aphididae BODY LENGTH 2–3mm SIMILAR SPECIES There are 600 species of aphid in Britain and even more in Europe, so take your pick.

JAN FEB MAR **APR MAY JUN JUL AUG SEP** OCT NOV DEC

COMMON GREEN LACEWING

Chrysoperla carnea

Of all the garden insects that might intrude indoors attracted by light or wafted in by happenstance, this must be one of the hardest to dislike or disdain, with its fragile appearance and enveloping cut-glass wings. In feeble flight it hardly seems to stir the air, while at rest the long antennae look like strands of silk. It is one of the garden's truly unmistakable visitors. While the adults do not appear to be capable of hurting a fly (they feed on nectar, pollen and honeydew from aphids), the larvae certainly are capable of doing so, and feed voraciously on aphids and other small animals. Adults sometimes overwinter in houses, turning autumn brown first.

WHERE TO FIND

Very common throughout the region, except in the far north.

EGG (several hundred, stalked, laid singly or in batches, on undersides of leaves) > **LARVA** (feeds on small insects) > **PUPA** (cocoon) > **ADULT**. Overwinters as adult, often in leaf litter. The life cycle may be completed in four weeks.

The larvae may attach the skins, body parts and debris of former meals to disguise themselves as they creep up on prey.

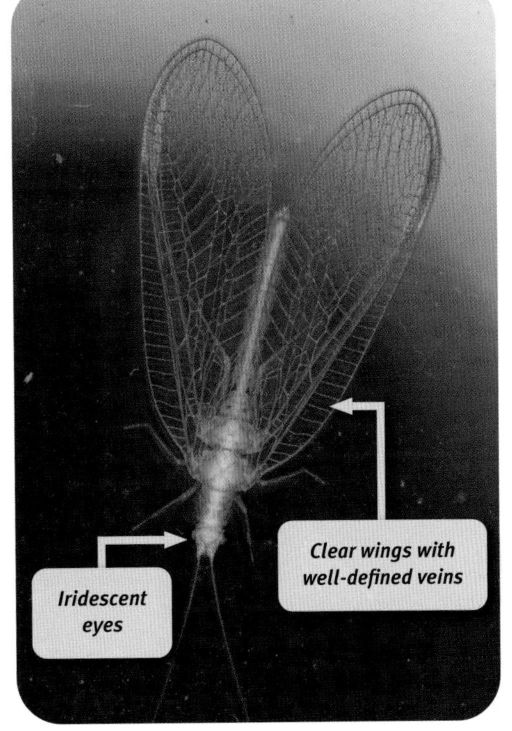

LEFT: *LARVA*; RIGHT: *ADULT*

FACT FILE

ORDER **Neuroptera** FAMILY **Chrysopidae** WING LENGTH **15mm** SIMILAR SPECIES Caddisflies (p. 100) are hairier, with fewer veins on the wings. There are about 12 very similar looking species, but they are otherwise unique.

JAN FEB MAR APR **MAY JUN JUL AUG SEP** OCT NOV DEC

BROWN LACEWING

Micromus variegatus

WHERE TO FIND

Common and widespread.

Much less known than their green relatives, Brown Lacewings are smaller and easily overlooked. They are usually secreted away in dense foliage, although they will break cover when attracted to light. The wings similarly envelop the body and are sublimely patterned, in this case with a network of spots and speckles. The antennae are as long as the body itself (not the wings). In contrast to the Common Green Lacewing (p. 43), the adults as well as the larvae eat insects such as aphids.

EGG (hundreds, on vegetation) > **LARVA** (feeds on small insects) > **PUPA** > **ADULT**. Overwinters as a pupa.

★ Brown Lacewing larvae never excrete; only in the adult stage do they do so.

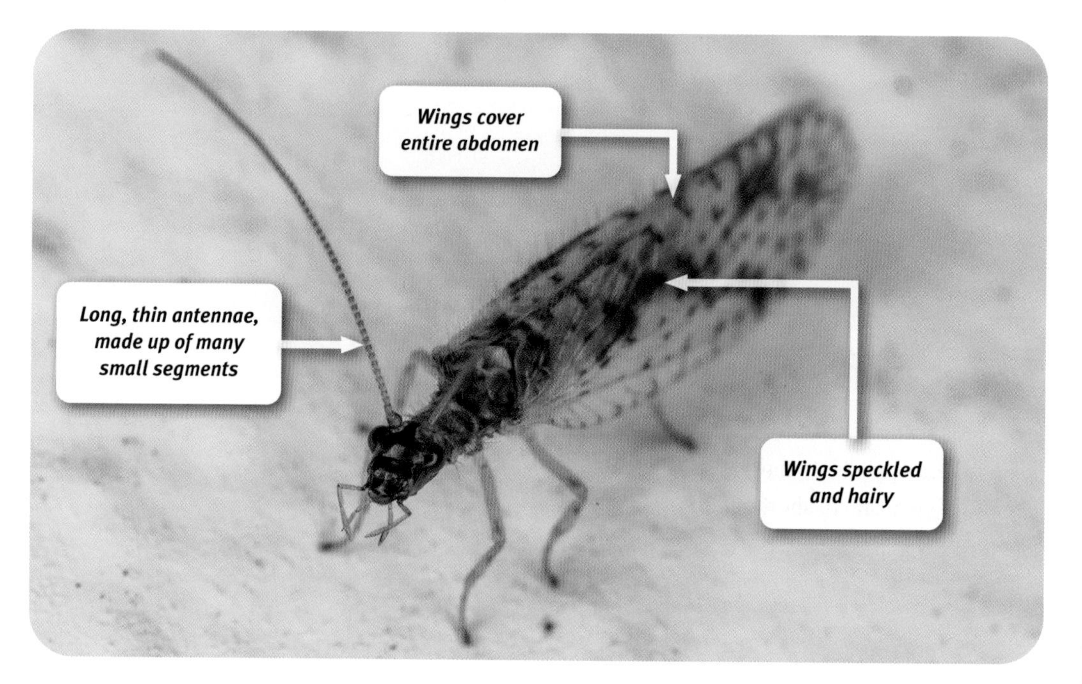

FACT FILE

ORDER Neuroptera FAMILY Hemerobiidae WING LENGTH 8mm SIMILAR SPECIES See caddisflies (p. 100). There are several species of brown lacewing that look very similar. The body plan is similar to that of a Common Green Lacewing but the colour is different.

JAN FEB MAR APR **MAY** **JUN** JUL AUG SEP OCT NOV DEC

ALDERFLY

Sialis lutaria

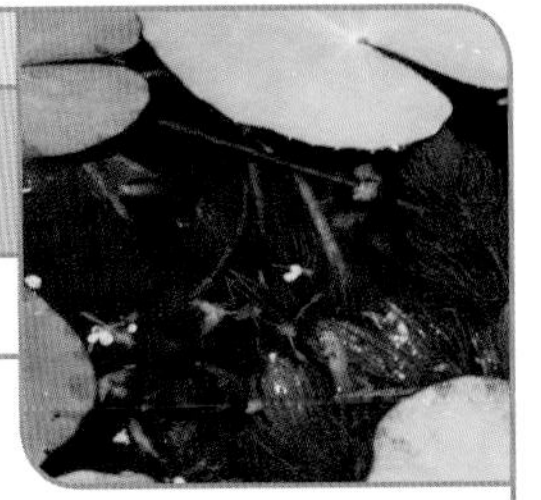

These unusual insects are found around garden ponds in the warm months of the year. They often perch on pondside vegetation, especially trees. They are quite large and fly like 'learner drivers', usually in the sunshine, and sometimes land on people, especially those wearing pale colours. The body is broad and dark with impressive long antennae and, as in lacewings, the wings envelop the body like a tent. In this case, though, the wings are olive-brown with dark veins, resembling a very expensive design. The adult stage only lasts for a few weeks, and most of an Alderfly's life is spent underwater.

WHERE TO FIND
Widespread and common in suitable habitats throughout the region, but rare in Ireland.

EGG (c. 500, laid on leaves or stones) > **LARVA** (enters water and lives in silt at the bottom, usually for two winters) > **PUPA** (in damp soil) > **ADULT**. Larvae are carnivorous, at first on micro-organisms and later on worms and insect larvae. Adults may not eat at all. Life cycle is usually two years.

The adults have an unfortunate tendency to fall into the water by mistake, shortening their 2–3-week lifespans.

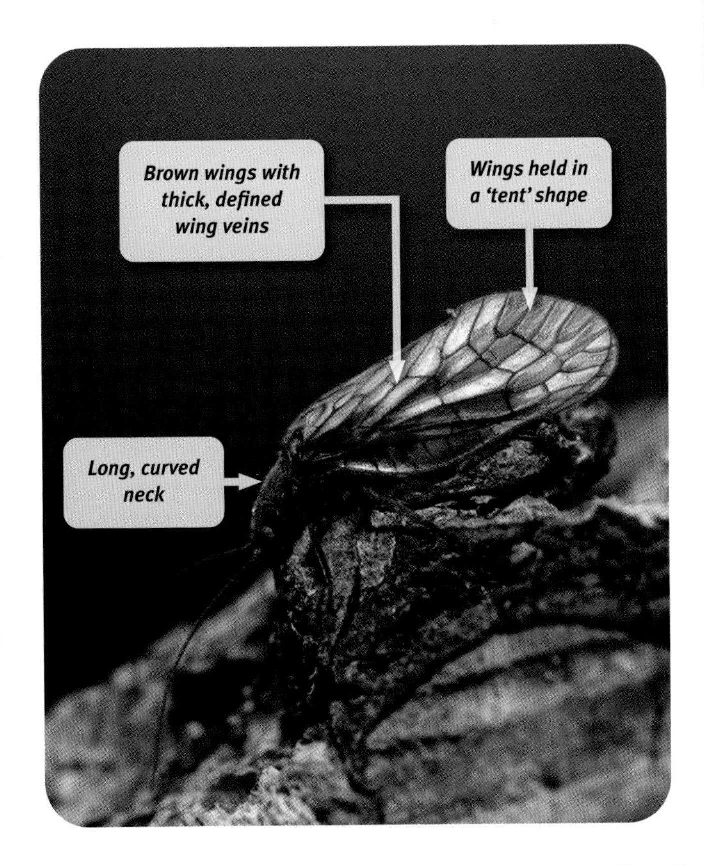

FACT FILE

ORDER Megaloptera FAMILY Sialidae BODY LENGTH 20mm WINGSPAN 22mm SIMILAR SPECIES Quite easy to identify. Has a noticeably long neck. Compare with caddisflies (p. 100).

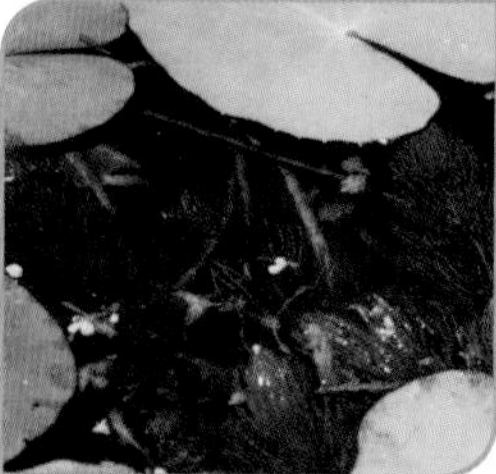

JAN | FEB | MAR | APR | MAY | JUN | JUL | AUG | SEP | OCT | NOV | DEC

COMMON WHIRLIGIG BEETLE

Gyrinus substriatus

WHERE TO FIND

Very common in garden ponds throughout the region.

Whirligig beetles are unmistakable. These are the miniature beetles that whirl and gyrate on the water's surface with wild restiveness, like aquatic headless chickens, their tiny elytra glinting as they 'swim' in rapid arcs. In the summer many of them gather in 'schools' and describe confusing, higgledy-piggledy patterns, an anti-predator tactic. The short lower pairs of legs are adapted for rowing and the front legs for grabbing invertebrates trapped on the surface. The antennae are specially adapted for detecting movements from both food and threats. They are the only beetles that live on the surface for most of their lives, although they can both dive and fly.

EGG (laid on aquatic vegetation) > **LARVA** (three instars, bottom dwelling in water) > **PUPA** (cocoon at water's edge) > **ADULT**. Larvae are carnivorous on small, bottom-dwelling worms and insects. Overwinters as an adult.

Whirligig beetles have two pairs of eyes, one above the other. One pair watches below the surface and the other above it.

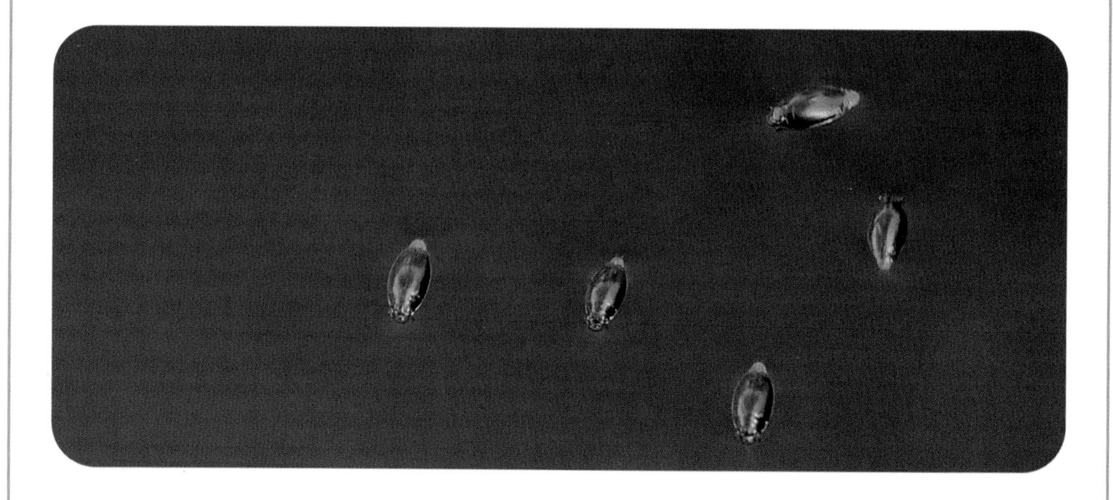

FACT FILE

ORDER Coleoptera FAMILY Gyrinidae BODY LENGTH 3–8mm SIMILAR SPECIES There are 12 species of whirligig beetle, all with small, oval bodies. Otherwise unmistakable.

JAN | FEB | **MAR** | **APR** | **MAY** | **JUN** | **JUL** | **AUG** | **SEP** | **OCT** | NOV | DEC

GREEN TIGER BEETLE

Cicindela campestris

One of Britain's most charismatic beetles, this gorgeous insect is a voracious predator of smaller invertebrates. It has the remarkable habit of leaping from the ground and making short ambush flights, snatching prey in mid-air before bringing it back to the ground. The grub-like larva is no less merciless, waiting concealed in a burrow for insects to come in range before summarily grabbing them. No other beetle has a similar green colour with a double necklace of pale spots. Note the very large eyes. This beetle is especially active in bright sunlight.

EGG > **LARVA** (makes its own burrow by excavation, up to 30cm deep, staying in it for up to two years) > **PUPA** (in burrow) > **ADULT**. Overwinters as a larva.

★ This beetle can run at 60cm per second.

WHERE TO FIND

A good find for a garden, but common in open, sandy areas and woodland rides throughout the UK and most of continental Europe, except for the very far north.

FACT FILE

ORDER Coleoptera FAMILY Cicindelidae BODY LENGTH 12–17mm SIMILAR SPECIES Unique colour and pattern.

JAN | FEB | MAR | APR | MAY | JUN | JUL | AUG | SEP | OCT | NOV | DEC

BLACK SNAIL BEETLE

Phosphuga atrata

WHERE TO FIND
Common and widespread throughout the region.

The six lithe legs of this beetle should give it a decidedly unfair advantage in catching up with its favourite foods, slugs and snails (mostly small ones). If eating the latter it climbs on top of the shell, bites the snail's head and releases an enzyme to dissolve the snail's tissues. In can then enter the shell and feast, using its peculiarly elongated neck. It also eats earthworms and, in keeping with being part of a family known as the carrion beetles, dead meat. On spring nights this tortoise-shaped, nocturnal beetle can be seen on pathways and in the grass, often in groups, but in hot weather it remains under logs and large stones, benefiting from sympathetic gardening.

EGG > **LARVA** (grub, predatory on snails and slugs) > **PUPA** (in soil) > **ADULT**. Overwinters as an adult.

It cannot fly and has to disperse by walking – obviously at more than a snail's pace.

FACT FILE

ORDER **Coleoptera** FAMILY **Silphidae** BODY LENGTH **10–16mm** SIMILAR SPECIES **The unusual rounded shape is distinctive.**

JAN | FEB | MAR | APR | MAY | JUN | JUL | AUG | SEP | OCT | NOV | DEC

SUN BEETLE

Amara aenea

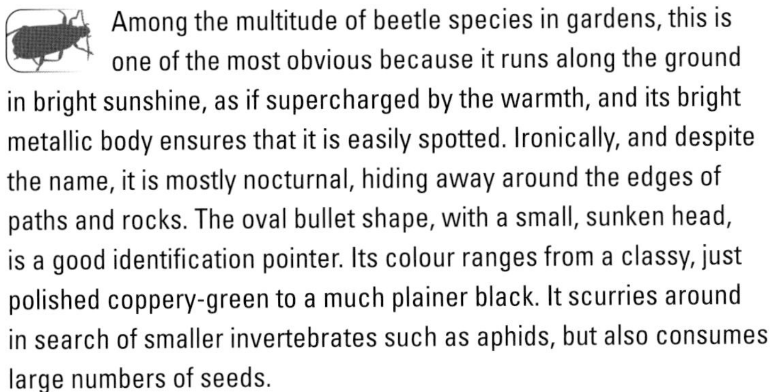

Among the multitude of beetle species in gardens, this is one of the most obvious because it runs along the ground in bright sunshine, as if supercharged by the warmth, and its bright metallic body ensures that it is easily spotted. Ironically, and despite the name, it is mostly nocturnal, hiding away around the edges of paths and rocks. The oval bullet shape, with a small, sunken head, is a good identification pointer. Its colour ranges from a classy, just polished coppery-green to a much plainer black. It scurries around in search of smaller invertebrates such as aphids, but also consumes large numbers of seeds.

WHERE TO FIND

Found throughout the region, but scarcer in Ireland and northern Scotland.

EGG (in dry soil) > **LARVA** (grub, in the ground) > **PUPA** (in the ground) > **ADULT**. Larvae and adults are omnivorous. Adults overwinter and may live for two years.

★ The larvae can be entirely carnivorous, entirely vegetarian or omnivorous.

FACT FILE

ORDER Coleoptera FAMILY Carabidae BODY LENGTH 6.5–9mm SIMILAR SPECIES The shape and strong sheen are distinctive, but there are many similar beetles.

JAN	FEB	MAR	**APR**	**MAY**	**JUN**	**JUL**	**AUG**	**SEP**	**OCT**	NOV	DEC

DEVIL'S COACH-HORSE

Ocypus olens

WHERE TO FIND
Widespread and common as far north as southern Scandinavia; rare in Ireland. Generally declining.

The body of this beetle seems to go on and on; it is remarkably long and slim. Its jet-black colour and nocturnal habits have helped to give rise to its association with all things evil in a number of cultures, hence the peculiar name. When threatened, it raises its abdomen up and over like the tail of a scorpion, an impressive display that is all for show. It can, however, also give a decent nip with its strong mandibles, as well as exude a chemical irritant. The jaws of this formidable beetle are used to pierce and crush small invertebrates such as worms, woodlice, slugs and spiders. They are chewed continually until the juices can be sucked out, leaving behind the inedible parts, such as exoskeletons. The larva lives in the soil, eating similar food, while the adults hide under stones and litter by day.

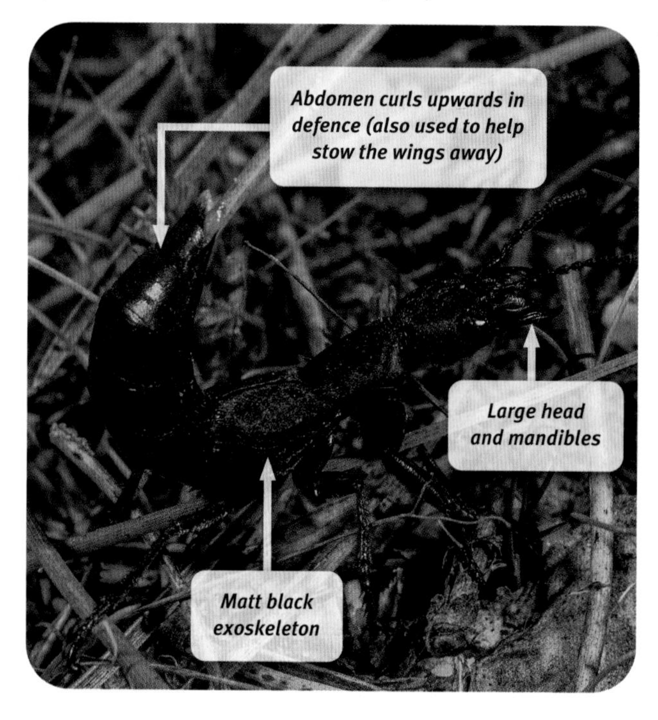

EGG (laid singly under moss or stones, in autumn) > **LARVA** (free-running, three instars, 150 days, in the ground) > **PUPA** (in the ground, 35 days) > **ADULT**. Larvae and adults are predators. Overwinters as a larva, but some adults may survive over the winter for a second season.

It often eats carrion, such as bugs or worms squashed on the surface of a path.

FACT FILE

ORDER **Coleoptera** FAMILY **Staphylinidae** BODY LENGTH **23–32mm** SIMILAR SPECIES **There are many species in the long-bodied rove beetle family, but this is one of the largest.**

JAN | FEB | MAR | APR | MAY | JUN | JUL | AUG | SEP | OCT | NOV | DEC

ROVE BEETLE

Tachyporus hypnorum

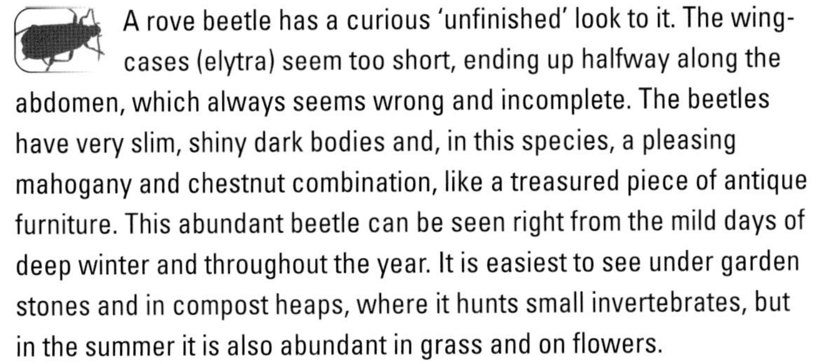

A rove beetle has a curious 'unfinished' look to it. The wing-cases (elytra) seem too short, ending up halfway along the abdomen, which always seems wrong and incomplete. The beetles have very slim, shiny dark bodies and, in this species, a pleasing mahogany and chestnut combination, like a treasured piece of antique furniture. This abundant beetle can be seen right from the mild days of deep winter and throughout the year. It is easiest to see under garden stones and in compost heaps, where it hunts small invertebrates, but in the summer it is also abundant in grass and on flowers.

WHERE TO FIND

Common and widespread throughout the region, mainly in lowlands.

EGG (in soil) > **LARVA** (grub, in soil, develops rapidly in three stages) > **PUPA** (in soil) > **ADULT** (in soil and litter) Overwinters as an adult. Both adults and larvae are predators, but larvae also consume seeds.

Venturing out on mild winter days depletes a beetle's food reserves. These beetles survive best during cold winters, when they remain inactive. The rove beetle family (Staphylinidae), with 63,000 species, comprises more species than any other family on the planet.

FACT FILE

ORDER **Coleoptera** FAMILY **Staphylinidae** BODY LENGTH **3–4mm** SIMILAR SPECIES There are multiple similar beetles in the same family. They bear a passing resemblance to a Common Earwig (p. 21).

JAN FEB MAR APR **MAY JUN JUL AUG** SEP OCT NOV DEC

STAG BEETLE

Lucanus cervus

WHERE TO FIND
Found throughout Europe, but only in the south of Sweden and Norway. Localized in south-east England, but common in some places, including London and the Home Counties. This animal has been declining for years and is greatly threatened by general tidying up in parks and gardens, removing old rotting stumps. It also falls prey to cats and dogs and is frequently run over.

One of the garden's true celebrities, this is our largest beetle. Only the males have the remarkably impressive mandibles, the equivalent of antlers, which are used in male-to-male clashes like the antlers of deer stags. Some individuals are endowed with much larger and more spiny antlers than others, and can use them to knock over an opponent or bring him to the ground – branches and logs are common battle grounds. On summer nights it often flies low and awkwardly and is attracted to lights.

EGG (laid in decaying wood, often at the roots and usually in an oak tree) > **LARVA** (grub, lives 4–7 years in wood) > **PUPA** (in soil) > **ADULT**. Larvae feed on wood and adults on sap.

One of the very first realistic depictions of any insect in Western art was a Stag Beetle painted by Albrecht Dürer in 1505. Stag Beetles are strongly attracted to ginger.

FACT FILE

ORDER Coleoptera FAMILY Lucanidae BODY LENGTH Male 35–75mm; female 28–45mm SIMILAR SPECIES The male's antlers are unique. It is larger than any other British beetle, and has a distinctive reddish hue to the wing-cases (elytra), lacking in the Lesser Stag Beetle (opposite).

JAN | FEB | MAR | APR | MAY | JUN | **JUL** | **AUG** | SEP | OCT | NOV | DEC

LESSER STAG BEETLE

Dorcus parallelipipedus

It might be 'the other' stag beetle, but this is still a very big, impressive beetle, quite a sight in a garden. Moreover, it is more common and more widely distributed than the Stag Beetle (opposite), in Britain reaching north as far as the Midlands. The male lacks the 'antlers' of the Stag Beetle and is always noticeably smaller. The body is more obviously parallel edged than that of the larger animal, lacks a reddish hue and, diagnostically, the front tibiae have a series of ridges running along them from front to back. It spends much of its time in rotting wood, but emerges at night, flying slowly, sometimes attracted to light. It is often found on tree trunks.

EGG (single, laid in tunnel in rotting wood, made by female) > **LARVA** (grub, three instars, lives 1–3 years in wood) > **PUPA** (in chamber in wood) > **ADULT**. Larvae feed on wood and adults on sap.

Adult Lesser Stag Beetles may live for 1–2 years, while their larger relatives die in their year of emergence.

WHERE TO FIND

Widespread and generally common, but declining in many places. In Britain, common in the southern half of England, including gardens and urban areas, but not found elsewhere. Found only in the south of Scandinavia.

FACT FILE

ORDER **Coleoptera** FAMILY **Lucanidae** BODY LENGTH **18–32mm** SIMILAR SPECIES **See Stag Beetle. No other British beetles are as big apart from the Great Diving Beetle *Dytiscus marginalis*.**

JAN FEB MAR **APR MAY JUN JUL AUG** SEP OCT NOV DEC

COMMON BROWN CLICK BEETLE

Athous haemorrhoidalis

WHERE TO FIND

Very common in lowlands throughout the region, but absent from Ireland.

This is a small brown beetle with a five-star trick – if handled, it can contort its body, make an audible click and leap into the air. This high entertainment is enabled by the beetle arching its back until a spine on the underside connects with a hinge that releases tension. It is a highly effective anti-predator shock mechanism, and also allows the insect to right itself. It additionally flies well. When not clicking, this is a super-common garden beetle that is often seen on white flowers with umbels, such as hogweed, as well as wild roses and nettles. Look for it and you will soon come across it. The factory-clean, matt-black head and neat, ridged, deep-maroon elytra (wing-cases) are also distinctive.

EGG (in soil) > **LARVA** (grub, 'wireworm', in soil, feeding on plant roots and some other insects, for two years) > **PUPA** (in soil) > **ADULT**. Overwinters as larva. Adults feed on pollen, nectar and young leaves.

The larvae often eat the caterpillars of winter moths.

FACT FILE

ORDER Coleoptera FAMILY Elateridae BODY LENGTH 10–15mm SIMILAR SPECIES There are 70 species of click beetle in the UK, and other small types of beetle also look similar.

JAN FEB MAR APR **MAY** **JUN** **JUL** AUG SEP OCT NOV DEC

COCKCHAFER

Melolontha melolontha

Often called the 'May Bug', this is very much an insect of warm nights in spring and early summer from dusk onwards. Solidly built, it is a slow crawler and an accident-prone flier, which progresses with an audible buzz as it careers into lighted windows. It often appears in swarms around tree tops and shrubs, landing to eat the leaves. The solid body appears to be clad in an old, worn-out overcoat, such is the chestnut colour with crinkled elytra with a floury lustre. Both sexes have stunningly stylish, fan-like antennae, with the male's being bigger and having seven sections, and the female's having six sections. There are several white triangular marks along the sides of the body. The season of this garden character is over all too soon.

WHERE TO FIND

Common and widespread in the region, but patchy in Scotland and only found in southern parts of Scandinavia.

EGG (c. 20, laid in roots among soil) > **LARVA** (grub, lives in soil for two years) > **PUPA** (from June, in soil) > **ADULT** (hatches in soil in late summer and overwinters there). Larvae eat roots in soil.

Fact: In the past, numbers used to hit plague-like peaks every 30 years or so.

FACT FILE

ORDER Coleoptera FAMILY Scarabaeidae BODY LENGTH 20–30mm SIMILAR SPECIES Apart from one or two uncommon chafer species, it is unmistakable.

JAN FEB MAR APR **MAY JUN JUL AUG SEP** OCT NOV DEC

ROSE CHAFER

Cetonia aurata

WHERE TO FIND
Widespread and often common north to southern Scandinavia. In Britain, found only in the south.

Most people's first thought on seeing the brilliantly metallic-green Rose Chafer on a sunny day is that it is a tropical insect that has been imported accidentally. However, this solidly built, strong-flying beetle is native to the southern UK and much of Europe, and feeds by day on pollen and nectar from a range of garden flowers, including elder, cow parsley, honeysuckles and, of course, roses. The extraordinary emerald-green sheen is so brilliant that it gives the impression that the insect's elytra has just been fiercely rubbed, and the transverse flaky-white markings suggest overzealous scraping.

EGG (batch laid in compost, leaf litter or rotting wood, June–July) > **LARVA** (large, fat grub with curved body, in compost for two years) > **PUPA** (in underground cell in compost or litter) > **ADULT** (hatches in autumn but remains underground until spring).

The Rose Chafer has unique mouthparts for taking up pollen, resembling a wet brush.

FACT FILE

ORDER Coleoptera FAMILY Scarabaeidae BODY LENGTH 14–21mm SIMILAR SPECIES The brightest of several similar species, most of which are uncommon.

JAN FEB MAR APR MAY **JUN** **JUL** **AUG** SEP OCT NOV DEC

COMMON RED SOLDIER BEETLE

Rhagonycha fulva

WHERE TO FIND
Very common throughout the region.

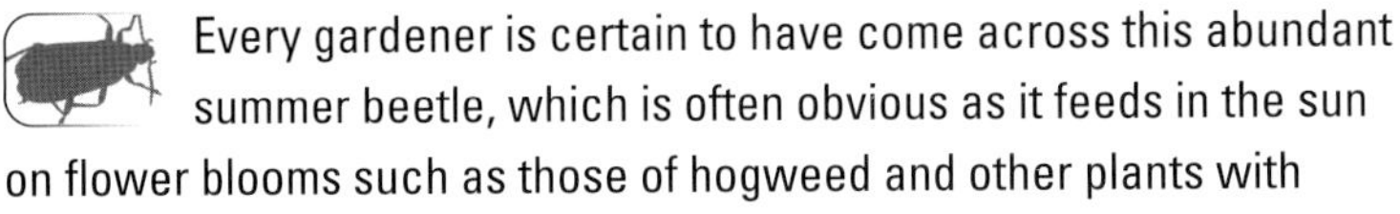

Every gardener is certain to have come across this abundant summer beetle, which is often obvious as it feeds in the sun on flower blooms such as those of hogweed and other plants with umbels (flat-topped or domed flowerheads arising from a plant's stem like umbrellas), thistles and ragworts. Several individuals, or even dozens, may visit a single flowerhead at any one time, especially on umbellifers. These are able fliers and can visit many flowers in a day. Although this species eats pollen and nectar, it is also a predator on smaller insects. Its habit of visiting flowers in daytime, its evenly narrow, tomato-sauce-coloured body, and the distinctive black tip to its elytra (wing-cases) all make it easy to identify.

EGG (batch laid in soil) > **LARVA** (grub, lives in leaf litter at base of grass) > **PUPA** (in soil in spring) > **ADULT**. The larvae eat many different invertebrates, even snails.

Some individuals of this species seem to spend much of their lives mating.

PAIR MATING

FACT FILE

ORDER Coleoptera FAMILY Cantharidae BODY LENGTH 7–10mm SIMILAR SPECIES There are many small red beetles, but they do not have the black 'wing-tip' or black antennae of this species.

JAN | FEB | MAR | **APR** | **MAY** | **JUN** | **JUL** | AUG | SEP | OCT | NOV | DEC

COMMON MALACHITE BEETLE

Malachius bipustulatus

WHERE TO FIND

Very common and widespread throughout the region.

This small, metallic green beetle looks as though it is wearing bright orange underpants, or as if its rear end is on fire. It is an abundant visitor to a wide range of garden blooms, where it feeds on pollen and nectar and, if the fancy takes it, insects. The adults appear in spring and are everywhere by midsummer. The males often climb to the tops of grass stems and hold up their antennae to attract females with chemicals.

EGG (in bark crevices and tussocks) > **LARVA** (grub, predatory, under bark) > **PUPA** (in soil) > **ADULT**. The larvae eat many different invertebrates, even slugs and snails.

The males may spend hours attracting females with pheromones. Some females tease males by approaching and sniffing them, but not mating.

FACT FILE

ORDER Coleoptera FAMILY Melyridae BODY LENGTH 5–8mm SIMILAR SPECIES There are many other colourful beetles of similar shape and size, but the red rear end in this one is distinctive.

JAN | FEB | MAR | APR | MAY | JUN | JUL | AUG | SEP | OCT | NOV | DEC

TWO-SPOT LADYBIRD

Adalia bipunctata

You only need to be able to count to two to identify this abundant ladybird. It is the only species with just two large spots on a red background. However, there are also many variations, some almost black, looking totally unlike the typical form. The legs are black. It is usually, but not always, found on the foliage of trees.

EGG > **LARVA** (with four growth stages) > **PUPA** (on leaf or hard surface, such as a fence) > **ADULT**. Larvae and adults are predators of aphids. Overwinters as an adult, often in the attics of houses.

A female only needs to mate once to produce many eggs for months afterwards – up to 500 in all.

WHERE TO FIND

Very common almost throughout the region, but has declined markedly in many areas, possibly due to competition with the Harlequin Ladybird (p. 63).

FACT FILE

ORDER Coleoptera FAMILY Coccinellidae BODY LENGTH 3–6mm
SIMILAR SPECIES Other ladybirds.

JAN | FEB | MAR | **APR** | **MAY** | **JUN** | **JUL** | **AUG** | **SEP** | **OCT** | **NOV** | DEC

CREAM-SPOT LADYBIRD

Calvia quattuordecimguttata

WHERE TO FIND

Widespread throughout the area.

This is obviously a ladybird, but without the usual black on red. Instead, this is a sartorial upgrade with 14 clear cream spots on a maroon background. The legs are brown. It is very common in gardens with deciduous trees, and found among foliage. This is a voracious predator, especially of plant lice (Psyllids), and a female may eat 50 or more in a single day.

EGG (each female may lay 300 eggs) > **LARVA** (with four growth stages) > **PUPA** > **ADULT**. Larvae and adults are predators, feeding on aphids and other small insects. Overwinters as an adult, in leaf litter or bark.

This ladybird is famous for covering its eggs with a powerful battery of waxy chemicals, which make them noxious and safe from predators.

FACT FILE

ORDER Coleoptera FAMILY Coccinellidae BODY LENGTH 4–5mm SIMILAR SPECIES Other ladybirds. There is a similar species called the Orange Ladybird *Halyzia sedecimguttata*, which has 12–16 spots and an orange hue. The Eighteen-spot Ladybird *Myrrha octodecimguttata* is also very similar, but has squarer spots.

JAN | FEB | MAR | APR | MAY | JUN | JUL | AUG | SEP | OCT | NOV | DEC

TWENTY-TWO-SPOT LADYBIRD

Psyllobora vigintiduopunctata

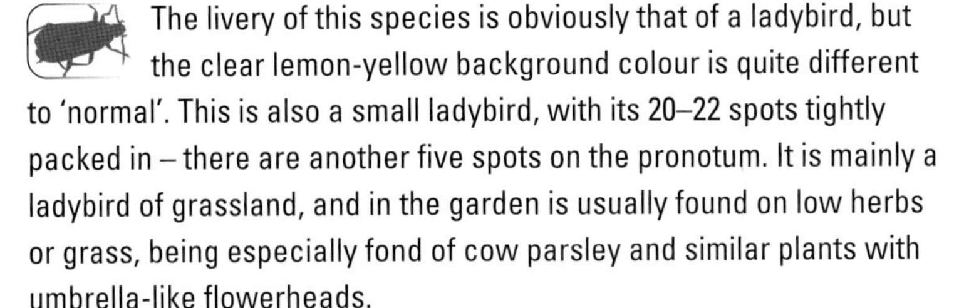

The livery of this species is obviously that of a ladybird, but the clear lemon-yellow background colour is quite different to 'normal'. This is also a small ladybird, with its 20–22 spots tightly packed in – there are another five spots on the pronotum. It is mainly a ladybird of grassland, and in the garden is usually found on low herbs or grass, being especially fond of cow parsley and similar plants with umbrella-like flowerheads.

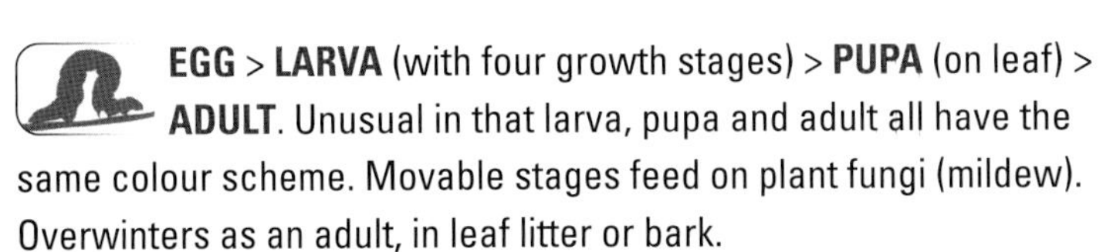

EGG > **LARVA** (with four growth stages) > **PUPA** (on leaf) > **ADULT**. Unusual in that larva, pupa and adult all have the same colour scheme. Movable stages feed on plant fungi (mildew). Overwinters as an adult, in leaf litter or bark.

It is often said that yellow ladybirds are harmful to humans but this is untrue – they can, however, be toxic to predators.

WHERE TO FIND

Found throughout the area.

FACT FILE

ORDER Coleoptera FAMILY Coccinellidae BODY LENGTH 3–4mm SIMILAR SPECIES Other ladybirds are generally red or orange.

JAN | FEB | MAR | APR | MAY | JUN | JUL | AUG | SEP | OCT | NOV | DEC

SEVEN-SPOT LADYBIRD

Coccinella septempunctata

WHERE TO FIND

Very common throughout much of the region apart from some northern areas. It has suffered a severe decline, probably due to competition with the Harlequin Ladybird (opposite).

The most common ladybird, this is one of the best known of all garden insects. It is popular not just for its appearance, but also for its ability to eat garden undesirables such as aphids. There are seven black, button-like spots on the scarlet wing-cases (elytra), the one by the head being centrally placed. The head is black with two large white patches. No other common ladybird has seven well-separated spots. It is found in any kind of low vegetation rather than in trees.

EGG > **LARVA** (with four growth stages) > **PUPA** (on leaf) > **ADULT**. Larvae and adults are predators, feeding on large numbers of aphids. Adults also feed on pollen and nectar. Overwinters as an adult, on dead leaves, seed heads, litter and similar.

The overall red colour and the size of the spots gives an indication to a predator of how toxic an individual is.

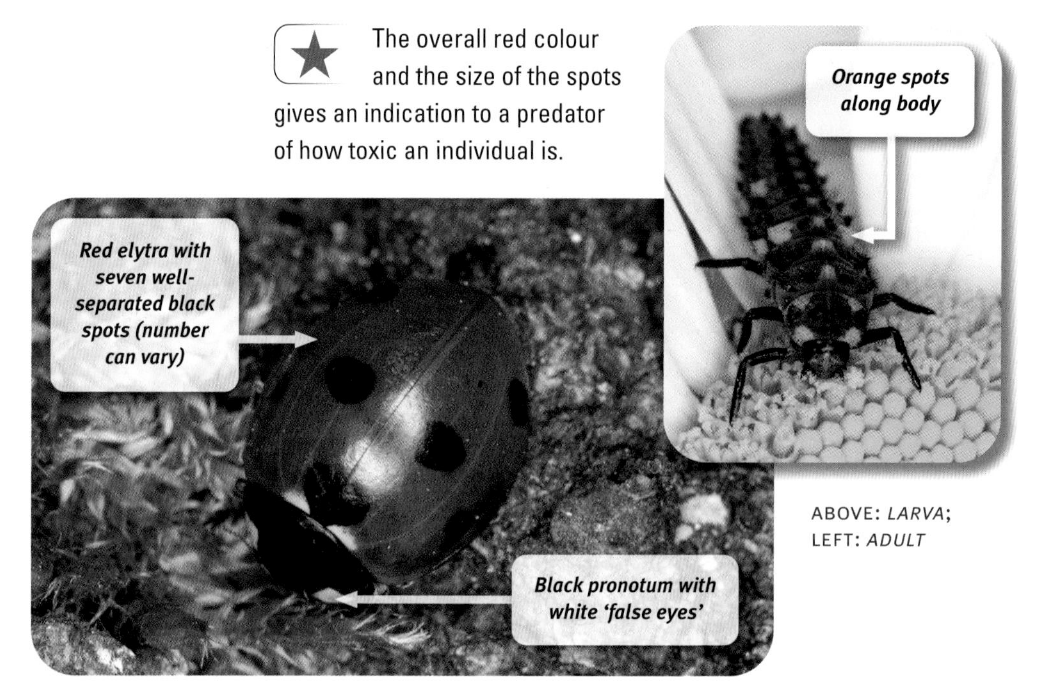

ABOVE: *LARVA*; LEFT: *ADULT*

FACT FILE

ORDER Coleoptera FAMILY Coccinellidae BODY LENGTH 5–8mm SIMILAR SPECIES Similar to many other ladybirds. The Harlequin Ladybird has many forms and may be easily confused with this species, but it usually has more than seven spots.

JAN | FEB | MAR | APR | MAY | JUN | JUL | AUG | SEP | OCT | NOV | DEC

HARLEQUIN LADYBIRD

Harmonia axyridis

This large ladybird usually has 16 spots on a red background, and a large amount of white on the pronotum. However, it is extraordinary variable and may look quite different, with more spots, or small spots, or with largely black elytra. A good feature is the rather tortoise-like, domed body. It is native to Asia and has been introduced to Europe. In the UK it appeared in 2004 and has spread widely and is now often the most common ladybird in urban settings. Its success has been detrimental to other species.

EGG > **LARVA** (with four growth stages) > **PUPA** > **ADULT**. Remarkably varied diet for a ladybird, eating pollen, nectar and even fruits. It has a marked predilection for flesh, too, eating copious aphids and, when these are not available, a wide range of other small creatures, including other ladybirds. Overwinters as adult, often in odd places such as compost bins, and indoors.

This species is a frequent cannibal in the larval stage. If the supply of aphids runs out, the larvae run amok among each other. The insects frequently fly high, up to 1,100m.

WHERE TO FIND

These days found throughout the region, having been introduced from east Asia. Its presence can be a bad sign for other ladybirds.

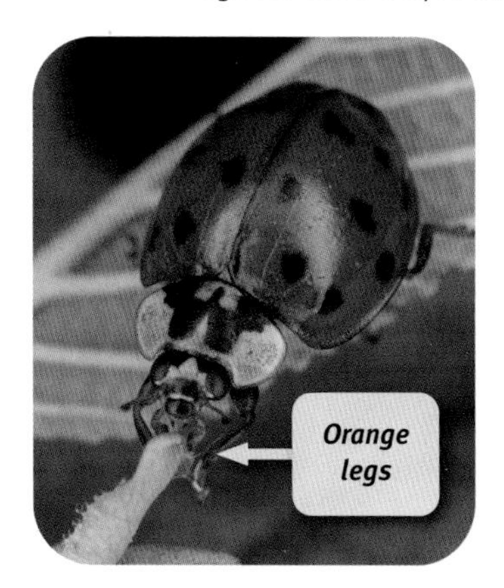

FACT FILE

ORDER Coleoptera FAMILY Coccinellidae BODY LENGTH 5–8mm SIMILAR SPECIES Often identified by size and domed appearance. Harlequins have a white spot/triangle on their head unlike other species, and orange legs. They look very similar to the Seven-spot Ladybird (opposite) but have more spots and the triangle marking.

JAN | FEB | MAR | **APR** | **MAY** | **JUN** | **JUL** | **AUG** | SEP | OCT | NOV | DEC

THICK-LEGGED FLOWER BEETLE

Oedemera nobilis

WHERE TO FIND

Mainly a southern species, absent from Scandinavia except Denmark, where it is rare. Increasingly common throughout England and Wales, and expanding north.

With its swollen hindlegs, the male of this beetle appears as though it has done a million squats in a gym. Both sexes seem to be wearing a waistcoat, or an ill-fitting jacket, as the elytra do not meet in the middle. This beauty lives up to its name, and is abundant on all sorts of different blooms, in both grassland and flowery gardens, where it leads a blameless life feeding on pollen. In common with many flower-loving beetles it is dazzlingly coloured, in this case a metallic green. The body is very long and thin. Look for it especially on daisies and bramble.

EGG (in stems) > **LARVA** (in dry stems of flowers) > **PUPA** (grub, in dry stems of flowers) > **ADULT**. Adults on flowers. Overwinters as a larva.

The function of the male's swollen limbs is entirely unknown.

FACT FILE

ORDER Coleoptera FAMILY Oedemeridae BODY LENGTH 8–11mm SIMILAR SPECIES There are many similar flower beetles. At a first glance, its body shape and long antennae resemble those of some longhorn beetles (opposite), but the size differences in both the body and the antennae are distinctive. The similar but less shiny *O. virescens* occurs further north.

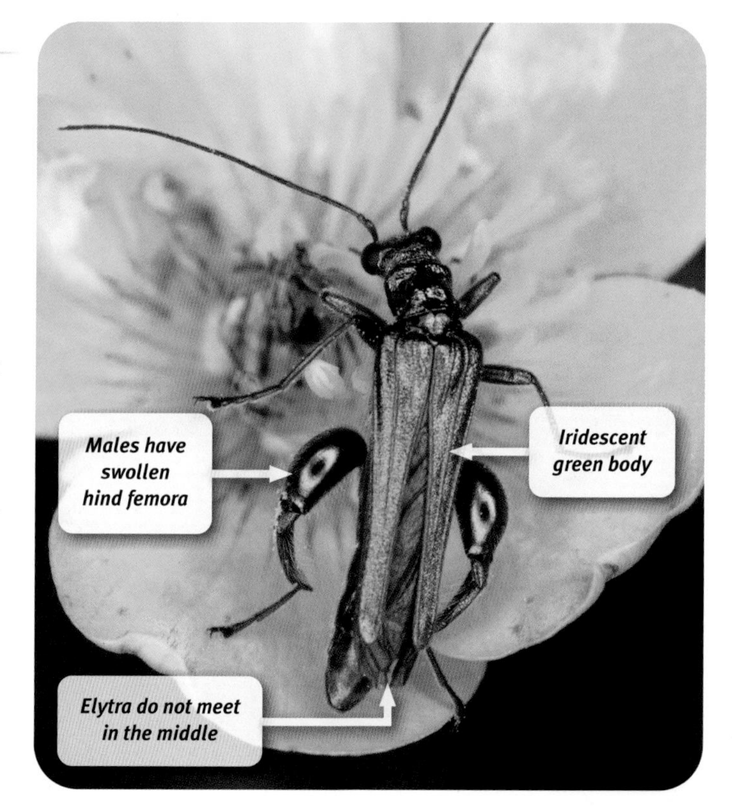

JAN FEB MAR APR **MAY JUN JUL AUG SEP** OCT NOV DEC

SPOTTED LONGHORN BEETLE

Rutpela maculata

This conspicuous garden beetle is long in every regard: it has extended antennae, which give rise to the name of longhorn beetles, long, spidery legs, and an elongated, slim body. The bright colour and black spots should also make it an easy insect to identify, although the pattern may vary strongly from what is shown here, sometimes occurring with a more orangey tinge and sometimes with more or fewer spots. It flies in bright sunshine and visits flower blooms of many kinds, from hawthorn to bramble to thistles, with umbels among the favourites. In common with several other beetles, it frequently mates on those same flowers. It requires the presence of rotting logs or branches for its larvae, so benefits when these are left in the garden.

WHERE TO FIND

Common over much of the region, but uncommon in Scotland, absent from Ireland and only present in southern Scandinavia.

EGG (in crevices in dead wood bark, often fence posts) > **LARVA** (in decaying wood) > **PUPA** (grub, in cell in wood) > **ADULT**. Adults eat mainly pollen; larvae eat wood. Overwinters as a larva, sometimes for two years.

This is a wasp mimic, which flies about in the day with the same bobbing, erratic action as a wasp.

FACT FILE

ORDER Coleoptera FAMILY Cerambycidae BODY LENGTH 13–20mm SIMILAR SPECIES Distinctive. Several other garden insects have a similar colour scheme. There are several similar species with a more northern distribution, such as the Four-banded Longhorn Beetle *Leptura quadrifasciata*, which reaches the Arctic.

WASP BEETLE

Clytus arietis

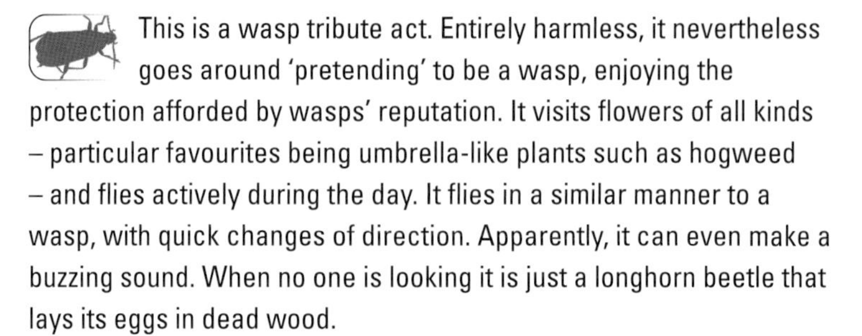

This is a wasp tribute act. Entirely harmless, it nevertheless goes around 'pretending' to be a wasp, enjoying the protection afforded by wasps' reputation. It visits flowers of all kinds – particular favourites being umbrella-like plants such as hogweed – and flies actively during the day. It flies in a similar manner to a wasp, with quick changes of direction. Apparently, it can even make a buzzing sound. When no one is looking it is just a longhorn beetle that lays its eggs in dead wood.

WHERE TO FIND

Common throughout the area north to southern Scandinavia and Scotland; rare in Ireland. A woodland insect, but often visits gardens.

EGG (in crevices in dead wood bark, often fence posts) > **LARVA** (tunnels into dead wood) > **PUPA** (grub, at end of tunnel) > **ADULT**. Adults eat mainly pollen; larvae eat wood. Overwinters as an adult.

When it lands on a log or trunk, it assumes the same jerky action as a wasp would when chewing wood for its nest construction.

Wasp mimic; yellow and black

Red legs

Hairy body

FACT FILE

ORDER Coleoptera FAMILY Cerambycidae BODY LENGTH 6–15mm SIMILAR SPECIES Wasps, of course. There are other beetles with black-and-yellow colouration.

JAN | FEB | MAR | APR | MAY | JUN | JUL | AUG | SEP | OCT | NOV | DEC

BLOODY-NOSED BEETLE

Timarcha tenebricosa

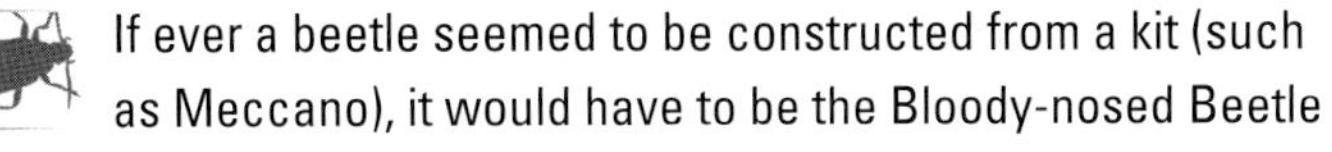

If ever a beetle seemed to be constructed from a kit (such as Meccano), it would have to be the Bloody-nosed Beetle – the legs and antennae have identikit segments, and the rounded black abdomen, head and thorax are so smart and neat that they look manufactured. Unusually, this beetle can be seen all year round, in various dry, open habitats, although it is a lucky garden that has it. It has a remarkable party trick that led to its name; if threatened or handled, it exudes – from the mouth, as it happens – drops of a red substance that looks remarkably like blood. The liquid smells unpleasant to would-be predators. This flightless beetle feeds on the stems and other foliage of bedstraws, living at ground level.

WHERE TO FIND

Common in the south of the region, not reaching Poland or Scandinavia, and also absent from Scotland and Ireland. Several similar species occur further north.

EGG (small batches covered with plant debris by female) > **LARVA** (in foliage; pupates after 5–8 weeks) > **PUPA** (grub, below ground, lasts a month) > **ADULT**. Larvae and adults eat leaves. Overwinters as an adult.

This beetle protects its eggs by regurgitating food over them.

FACT FILE

ORDER Coleoptera FAMILY Chrysomelidae BODY LENGTH 10–20mm SIMILAR SPECIES More rotund and slower than many similar-sized dark beetles.

JAN | FEB | MAR | APR | MAY | JUN | JUL | AUG | SEP | OCT | NOV | DEC

ROSEMARY BEETLE

Chrysolina americana

WHERE TO FIND

Common on the Continent except in Scandinavia, and in the UK expanding fast from southern England.

There is no doubting the beauty of this ladybird-sized beetle. The extraordinary rainbow-like streaks of iridescent colours are simply stunning, and there are also longitudinal double rows of pockmarks, reminiscent of the pattern of a sea-urchin skeleton. It is a native of southern Europe that has been steadily colonizing Britain since being accidentally introduced in the 1960s. After a slow expansion it has exploded in recent years and has now reached southern Scotland. It feeds, as its name implies, on rosemary, as well as on thyme, sage and lavender.

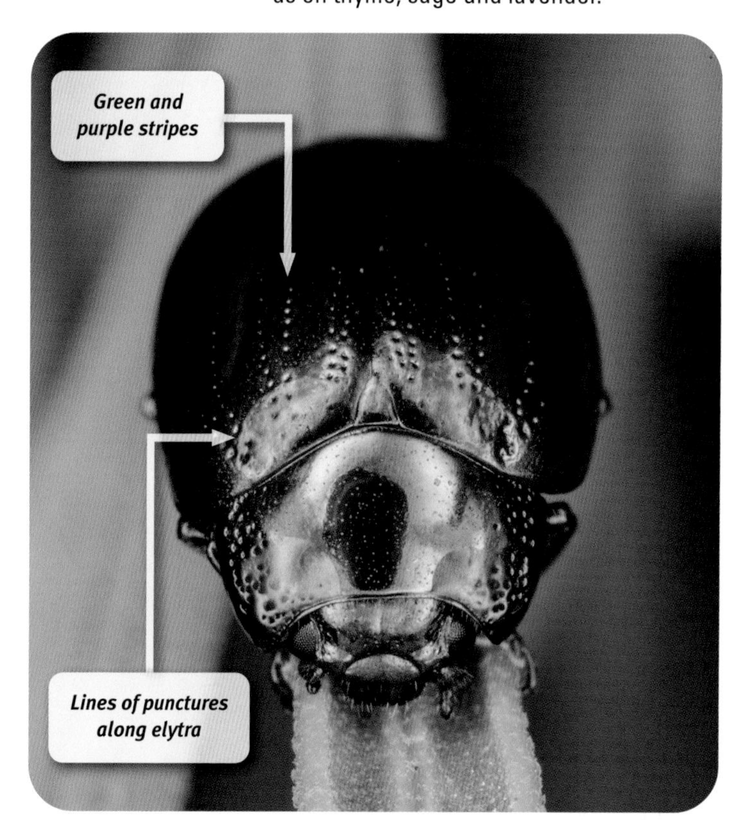

EGG (laid on leaves of host plant) > **LARVA** (grub, in soil) > **PUPA** (in the same cell in soil) > **ADULT**. Larvae and adults eat leaves. Overwinters as a larva.

When males are mating, they hold on very tight – equivalent to 45 times their own body weight.

FACT FILE

ORDER Coleoptera FAMILY Chrysomelidae BODY LENGTH 6–8mm SIMILAR SPECIES Some other beetles are multicoloured and shiny, but the pattern is unique. There are many similar species, some of which, such as *C. polita*, occur further north.

JAN FEB MAR APR **MAY JUN JUL AUG** SEP OCT NOV DEC

LILY BEETLE

Lilioceris lilii

At first sight you might mistake the Lily Beetle for a spotless ladybird, but do not be fooled – this beetle means trouble. It is a significant pest of lilies and fritillary flowers in gardens, nurseries and garden centres, leaving holes in all parts of the flower, including the blooms. It is undeniably attractive, though, with its intense carmine-red back (thorax and abdomen) and jet-black legs and antennae. The overall body shape is somewhat tortoise-like. It flies from garden to garden, looking for its food plants, almost all summer.

EGG (up to 400 per female, on lily leaves; batches of 10 or so) > **LARVA** (grub that accumulates detritus around it for protection) > **PUPA** (in soil within silk) > **ADULT**. Adults and larvae eat leaves, stems and even flowers of lilies. Overwinters as an adult in soil.

This beetle can squeak when handled or otherwise disturbed. The larva covers itself with its own dung for protection.

WHERE TO FIND

Introduced from Asia, and now common in temperate parts of the region, and spreading north.

FACT FILE

ORDER Coleoptera FAMILY Chrysomelidae BODY LENGTH 6–8mm SIMILAR SPECIES Ladybirds. There are other small, bright red beetles, but this one is broader bodied than most.

JAN | FEB | MAR | APR | MAY | JUN | JUL | AUG | SEP | OCT | NOV | DEC

GREEN DOCK BEETLE

Gastrophysa viridula

WHERE TO FIND
Very common throughout the region north to southern Scandinavia.

At first glance you can hardly distinguish this shining green beetle from the Mint Leaf Beetle (opposite), but the point is that this one is found on dock leaves, while its relative occurs only on mints. It is common among many British insects to have very specific requirements, and this is a good garden example. This beetle is abundant on dock and sorrel throughout the region and can easily be found in summer on leaves and stems.

EGG (batches of up to 60, on host leaves) > **LARVA** (grub, on leaves, sometimes defoliating plants, three instars) > **PUPA** (in soil, 10 days) > **ADULT**. Larvae and adults eat dock leaves. Overwinters as an adult, in tussocks or litter.

If the food plants run out, the larvae eat their own eggs and other larvae.

FACT FILE

ORDER Coleoptera FAMILY Chrysomelidae BODY LENGTH 4–8mm SIMILAR SPECIES The Mint Leaf Beetle is larger.

JAN FEB **MAR APR MAY JUN JUL AUG SEP OCT** NOV DEC

MINT LEAF BEETLE

Chrysolina herbacea

This gorgeous emerald-green, shiny beetle is strongly associated with various species of mint, including those that grow near or in the water, so it is frequently found in the vicinity of garden ponds. Where it occurs, it is often common among the foliage. It is always, too, in mint condition, a slow-moving but well-proportioned beetle with a stunning metallic sheen. Close up, the body can be clearly seen to be pockmarked with small dimples.

EGG (laid on leaves of host plant in May and June) > **LARVA** (grub, in soil) > **PUPA** > **ADULT**. Larvae and adults eat leaves of host plant. Overwinters as a larva, but some adults survive into autumn and winter, and some overwinter in thick vegetation.

The saliva of this beetle contains toxins that kill the mint plant.

WHERE TO FIND

Common north as far as southern England. Absent from Scandinavia, Scotland and Ireland.

FACT FILE

ORDER Coleoptera FAMILY Chrysomelidae BODY LENGTH 7–11mm SIMILAR SPECIES There are a number of other shiny green beetles, but this one is relatively large and is usually on the host plant. Several similar species occur further north.

JAN FEB MAR **APR MAY JUN JUL** AUG SEP OCT NOV DEC

GREEN NETTLE WEEVIL

Phyllobius pomaceus

WHERE TO FIND

Occurs throughout the region into southern Scandinavia; scattered records further north.

The weevils are a group of beetles with 'long noses'. There are many species in the garden, often hard to identify, but one of the bigger, most conspicuous ones is the Green Nettle Weevil. Its greenish colouration varies due to a covering of scales that are easily rubbed off by wear and tear, so that some individuals are almost black. In weevil style, the antennae can be flexed halfway along, like elbows. Many weevils are very attached to their hosts and this one is no exception – it is always found around nettles, often in large numbers from May onwards. It scuttles in a spider-like way, almost robotically. With its big eyes, it looks as though it is wearing goggles.

EGG (laid in soil below nettles) > **LARVA** (around roots of nettles) > **PUPA** (grub, in soil) > **ADULT**. Adults eat foliage, the larvae roots. Overwinters as an adult.

There are more than 60,000 species of weevil in the world, more than all the birds, mammals, reptiles, amphibians and fish combined.

FACT FILE

ORDER Coleoptera FAMILY Curculionidae BODY LENGTH 7–10mm SIMILAR SPECIES There are many other similar species of weevil, but this one is quite distinctive.

JAN FEB **MAR APR MAY JUN JUL AUG SEP OCT** NOV DEC

PEA WEEVIL

Sitona lineatus

You will see the signs of activity before encountering the Pea Weevil. The adults chew leaf edges and make easily recognized, semi-circular notches in the foliage of peas, beans, clover and other related plants. This species is sometimes troublesome, both as an adult and as a larva, the latter chewing roots. It abounds in the foliage of legumes in the garden in summer, and if you disturb one it will drop to the ground and play dead. In winter, the adults all 'migrate' to stacks of straw or grass and are generally inactive. For recognition purposes, the food plant is a guide, and this beetle also has the typical 'snout' of a weevil and neat stripes along its back.

WHERE TO FIND

Common north to southern Scandinavia, but absent from Scotland.

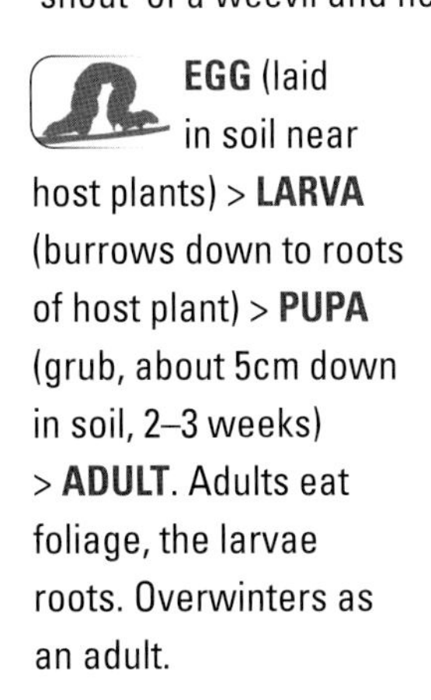

EGG (laid in soil near host plants) > **LARVA** (burrows down to roots of host plant) > **PUPA** (grub, about 5cm down in soil, 2–3 weeks) > **ADULT**. Adults eat foliage, the larvae roots. Overwinters as an adult.

Females of this species can be incredibly productive: they can lay 1,600 eggs each. In early spring, males send out a chemical to attract others to suitable plants.

FACT FILE

ORDER Coleoptera FAMILY Curculionidae BODY LENGTH 3.5–5.5mm SIMILAR SPECIES There are many other weevil species of similar colour, but the stripes on this one are distinctive.

JAN FEB MAR **APR MAY JUN JUL AUG SEP** OCT NOV DEC

CRANEFLIES

Tipulidae spp.

WHERE TO FIND

Very common throughout the region.

Craneflies are better known by the name 'daddy longlegs' and they are the broken-down car wreck among garden insects. They fly as if steering had never been invented, veering wildly like learner drivers, and often barely managing to take off from a garden lawn at all, apparently dragging their over-long legs over the grass. The legs commonly break off, but they fly on regardless, often fatally attracted to light and flying against windows and walls. To look at they resemble giant mosquitoes, but those ominous extended mouthparts are too weak to harm us, and the adults often do not eat much anyway. The larvae, the leatherjackets, chew away at the roots of grasses and leave yellow patches in lawns.

EGG (laid on damp soil) > **LARVA** (elongated grub, in soil, feeds on roots and stems of many plants) > **PUPA** > **ADULT**. Adults live for 10–14 days.

In 1935 leatherjackets were so abundant at Lord's Cricket Ground in London that matches had to be abandoned, and the wicket became a nightmare for those batting.

FACT FILE

ORDER Diptera FAMILY Tipulidae WING LENGTH 15–30mm SIMILAR SPECIES There are several cranefly species in gardens.

JAN FEB MAR **APR** **MAY** **JUN** JUL AUG SEP OCT NOV DEC

ST MARK'S FLY

Bibio marci

This is a marvellously jet-black, smart, hairy and leggy fly that spends much time in mating swarms, hovering up and down at human head height, where it is easy to see. It is mainly a woodland fly but visits gardens. At rest the clear wings fold neatly over the body and the legs are splayed. In flight these very long legs dangle and the abdomen is raised, giving a characteristic shape. The eyes are very large in the male, but smaller in the female. The flies have a 'messy flight' – looking almost as if they are intoxicated.

EGG (batches laid in rotting vegetation) > **LARVA** > **PUPA** > **ADULT** (emerges around the time of St Mark's Day, 25 April, hence the name).

The eye of the male is divided into two by a groove. The top half is used to look for females while hovering, while the bottom half is used to monitor the fly's position above the ground.

WHERE TO FIND

Common and widespread throughout the region.

FACT FILE

ORDER Diptera FAMILY Bibionidae BODY LENGTH 10–15mm SIMILAR SPECIES Much larger than gnats or mosquitoes. There are several similar closely related species.

JAN | FEB | MAR | APR | MAY | JUN | **JUL** | **AUG** | SEP | OCT | NOV | DEC

DARK-WINGED FUNGUS GNAT

Sciara hemerobioides

WHERE TO FIND
Widespread and fairly common throughout the region, but easily overlooked.

It might seem odd to call a gnat handsome, but if any gnat deserves the accolade it would have to be this one. Jet black, with a yellow abdomen, it is a tiny insect that would remain under the radar if it did not sometimes emerge from the soil of indoor potted plants. The adults are visitors to flowers, feeding on the nectar of umbellifers such as hogweed, but their size keeps their profile low. The name refers to the behaviour of the larvae, which feed among the underground filaments of fungi.

EGG (in soil among fungi) > **LARVA** (among mycelia; may also feed on decaying wood) > **PUPA** (in soil) > **ADULT**.

Despite their size, they are very tough and can live in extreme habitats above 4,000m.

FACT FILE

ORDER Diptera FAMILY Sciaridae BODY LENGTH 5–6mm SIMILAR SPECIES There are many other gnats and small flies, but none have the colour combination of this one.

JAN FEB MAR APR MAY JUN JUL AUG SEP OCT NOV DEC

MOTH FLIES

Psychodidae spp.

There are few compensations for cleaning drains and sinks or dealing with rubbish bins, but bumping into this wonderful insect is at least one of them. Small clouds of them may flit around dirty, damp corners of the household and garden, but they are perfectly harmless. Also known as 'owl midges', they look like tiny moths but have only one pair of wings.

EGG (laid in stagnant water and slime) > **LARVA** (feeds on detritus) > **PUPA** > **ADULT** (lives about 20 days).

WHERE TO FIND
Common and widespread throughout the region.

FACT FILE

ORDER **Diptera** FAMILY **Psychodidae** BODY LENGTH **2–4mm** SIMILAR SPECIES **Unique.**

JAN FEB MAR **APR MAY JUN JUL AUG SEP** OCT NOV DEC

NON-BITING MIDGES

Chironomus spp.

Few insects cause more panic for absolutely no reason than non-biting midges. The name tells the story; they do not eat blood. However, they often appear in clouds and settle on people, which can be disconcerting.

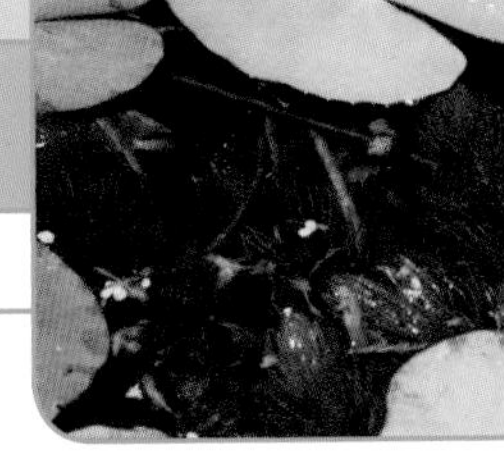

WHERE TO FIND
Abundant anywhere near water.

EGG (spiral strings attached to plants) > **LARVA** (aquatic, bright red, worm-like, feeds on organic matter, and lives in tube it constructs out of mud and silk) > **PUPA** (in mud) > **ADULT**.

FACT FILE

ORDER **Diptera** FAMILY **Chironomidae** BODY LENGTH **10mm** SIMILAR SPECIES **There are many types of midge, but the humped thorax of a non-biting midge is distinctive.**

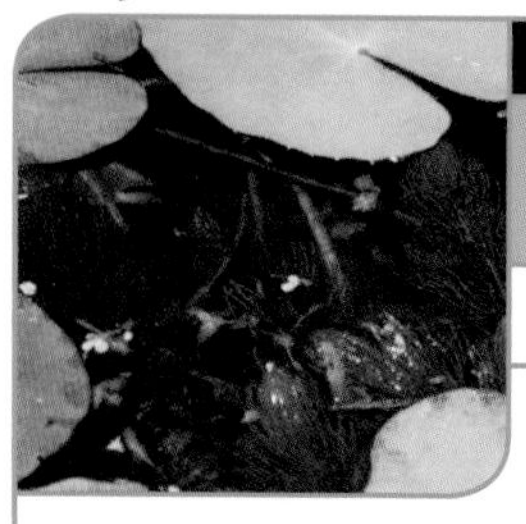

JAN | FEB | MAR | APR | MAY | JUN | JUL | AUG | SEP | OCT | NOV | DEC

MOSQUITOES

Culicidae spp.

WHERE TO FIND

Abundant everywhere.

This is one garden insect, at least, that approaches humans enthusiastically. The itch is part of the wildlife experience of every summer. However, without mosquitoes there would be fewer avians in the sky, because birds such as swifts and house martins consume large numbers and depend on these famous flies. Moreover, by no means does every species bite humans and even among those that do, only the females feed on blood, which they need for fertilizing their eggs. Males lack piercing mandibles and instead feed on nectar. Mosquitoes are bigger than midges, with very long, slender legs and a long proboscis.

EGGS (laid on the water's surface, glued together to form rafts) > **LARVA** (aquatic, moves in jerky motion) > **PUPA** (comma shaped, moves around) > **ADULT**.

Some mosquitoes beat their wings at nearly 1,000 times a second.

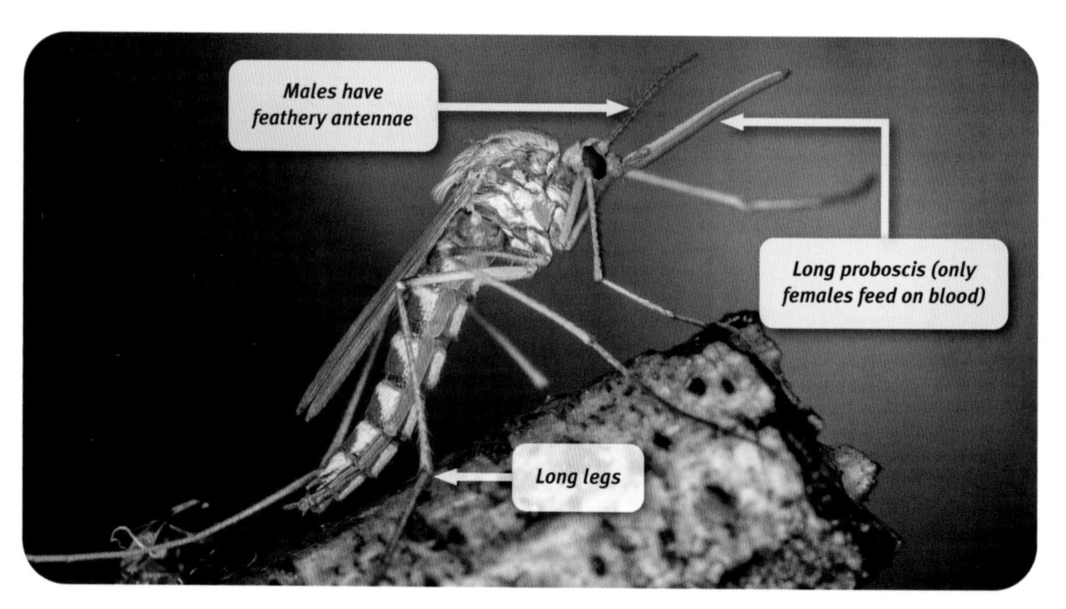

CULISETA ANNULATA

FACT FILE

ORDER **Diptera** FAMILY **Culicidae** BODY LENGTH **6–7mm** SIMILAR SPECIES **There are several species of mosquito and many hundreds of gnats, all of which look similar.**

JAN	FEB	MAR	APR	**MAY**	**JUN**	**JUL**	**AUG**	SEP	OCT	NOV	DEC

HORSEFLIES

Tabanidae spp.

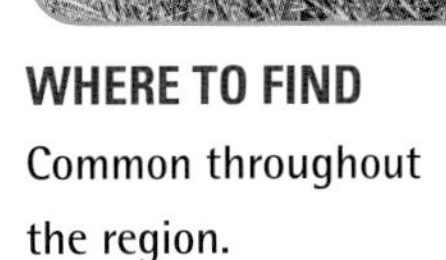

These are unpopular insects. As their name implies, horseflies can bite through the hide of a horse or cow, so they have no difficulty puncturing our skin – and it hurts. The impressive mouthparts of the females are effectively stabbing devices with sharpened edges, so these flies are actually 'pool feeders', making the wound first, then lapping up the blood. Only the females lap up blood, which they need to provide enough protein for their eggs. Both sexes also feed on nectar – and indeed are excellent pollinators. These are large flies, with an unexpected mollifying feature: their eyes are stunning, with psychedelic bands of colour.

WHERE TO FIND

Common throughout the region.

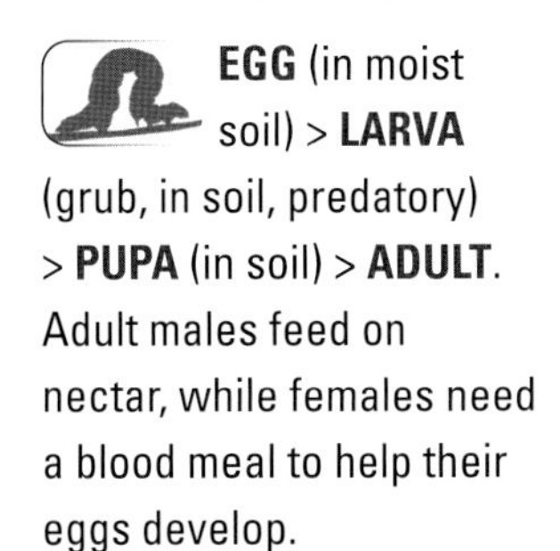

EGG (in moist soil) > **LARVA** (grub, in soil, predatory) > **PUPA** (in soil) > **ADULT**. Adult males feed on nectar, while females need a blood meal to help their eggs develop.

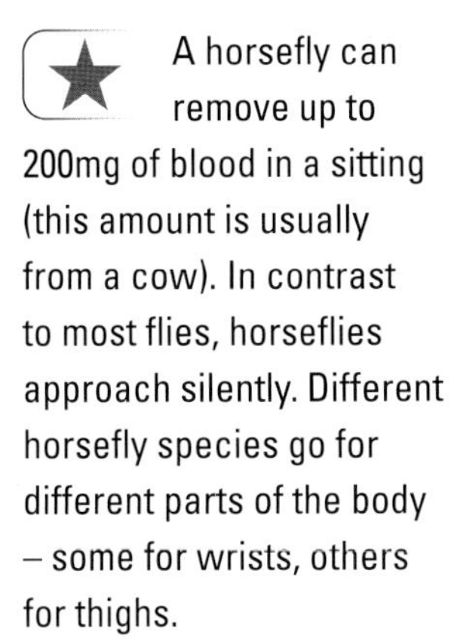

A horsefly can remove up to 200mg of blood in a sitting (this amount is usually from a cow). In contrast to most flies, horseflies approach silently. Different horsefly species go for different parts of the body – some for wrists, others for thighs.

Hairy-legged Horsefly Hybomitra bimaculata

FACT FILE

ORDER Diptera FAMILY Tabanidae BODY LENGTH 13–16.5mm SIMILAR SPECIES There are 30 horsefly species in Britain, many more in Europe. Most are larger than typical flies.

JAN FEB MAR APR **MAY JUN JUL AUG** SEP OCT NOV DEC

BROAD CENTURION

Chloromyia formosa

WHERE TO FIND
Very common in lowlands throughout the region.

Some flies seem to spend much of their lives simply sitting on a leaf and lazing in the sun. A good example is the Broad Centurion, a member of a lethargic group called the soldierflies, named after their metallic, glinting 'body armour', which is often colourful. This one spends much time in gardens, usually low down, and it is partial, as is many a garden insect, to the blooms known as umbellifers, with florets in an umbrella shape. The male and female Broad Centurion are quite different, the male sporting a green thorax and gold abdomen, the female a gold thorax and royal blue abdomen.

EGG > **LARVA** (woodlouse-like grub, in decomposing wood or leaves in soil) > **PUPA** (in soil) > **ADULT**. Adults feed on nectar. Overwinters as a larva.

★ Soldierflies appear to have an irresistible attraction to rotted grapefruit peel.

FACT FILE

ORDER Diptera FAMILY Stratiomyidae BODY LENGTH 8–9mm SIMILAR SPECIES Greenbottle (p. 97) and many other flies.

JAN FEB **MAR APR MAY JUN** JUL AUG SEP OCT NOV DEC

DARK-EDGED BEE-FLY

Bombylius major

This is possibly the most charismatic garden insect that most people have never heard of. It hovers in front of spring flowers with its long 'nose' (proboscis) protruding, like a miniature hummingbird. The body is dark and furry, and it has very long legs. It is fond of feeding at primroses. Its lifestyle is also remarkable, since it is a parasite on solitary bees (mainly *Andrena* species).

EGG (adult females literally throw their eggs towards the hole of a solitary bee, using a flicking action of the body in flight) > **LARVA** (three instars) > **PUPA** > **ADULT**. The larvae feed on bee grubs and their food. Hibernates as a pupa inside the nest hole.

Bee-flies have a type of flight that is quite unique. They can 'jaw' – rotate at high speed around a vertical axis. Nobody knows why.

WHERE TO FIND

Fairly common in sheltered gardens throughout temperate parts of the region, but rare in Ireland.

ABOVE: *LATERAL VIEW*; TOP: *NECTARING*

FACT FILE

ORDER Diptera FAMILY Bombyliidae BODY LENGTH 6–12.5mm SIMILAR SPECIES These insects resemble tiny bumblebees – mimicking is a good skill to have when acting as a parasite to solitary bee nests. There are several similar species.

JAN | FEB | MAR | APR | MAY | JUN | JUL | AUG | SEP | OCT | NOV | DEC

MARMALADE HOVERFLY

Episyrphus balteatus

WHERE TO FIND

Very common throughout the region.

This small, delicate hoverfly is superficially bee- or wasp-like but is able, like other hoverflies, to 'hang in the air' on super-fast wingbeats. This is a great example of what is known as 'Batesian mimicry', where a harmless and edible species dresses up as a harmful species. It has a unique pattern of twin dark bands, one thick and one thin, separated by orange bands. It feeds at flower blooms. Some individuals are darker than others, a smart chocolate-brown.

EGG > **LARVA** (grub, three stages of growth) > **PUPA** > **ADULT**. The larvae feed on aphids. Hibernates as an adult.

In Britain in late summer it may arrive in swarms of millions, wafted over on fair winds from the Continent. Up to four billion individuals may be involved.

ABOVE: *FEMALE*; LEFT: *MALE*

FACT FILE

ORDER Diptera FAMILY Syrphidae WING LENGTH 6–10mm SIMILAR SPECIES Other hoverflies, bees, wasps and sawflies.

JAN FEB MAR **APR MAY JUN JUL AUG SEP OCT NOV** DEC

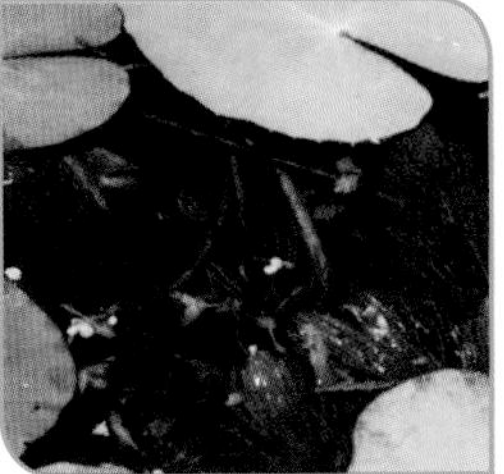

TIGER HOVERFLY

Helophilus pendulus

This chunky hoverfly has black-and-yellow stripes on the thorax at right angles to those on the abdomen. The species is particularly common around garden ponds, where it frolics and lounges around floating vegetation, buzzing noisily. It also visits flowers far from water.

WHERE TO FIND
Very common throughout the region.

EGG > **LARVA** (in wet, decaying vegetation) > **PUPA** > **ADULT**. The larvae often live in leaves at the edges of ponds, filter-feeding on micro-organisms.

The larvae are aquatic and have a long breathing tube, like a snorkel.

FACT FILE

ORDER Diptera FAMILY Syrphidae WING LENGTH 8.5–11mm SIMILAR SPECIES Other hoverflies, bees, wasps and sawflies.

JAN FEB MAR APR **MAY JUN JUL AUG SEP OCT** NOV DEC

BATMAN HOVERFLY

Myathropa florea

WHERE TO FIND

Very common throughout the region.

Every summer, Batman visits gardens in the guise of a hoverfly, given away only by the extraordinary pattern on the thorax, a cleverly concealed Batman logo, set against a bright, shimmering, lemony colouration. Otherwise, this is a large hoverfly species with a distinct resemblance to a bee or wasp that offers good protection from predators.

EGG > **LARVA** (in rot-holes in trees) > **PUPA** > **ADULT**. The larvae filter feed on micro-organisms.

When a branch breaks off a tree, the scar often fills with water and vegetation, which then decays into a smelly ooze. This is the secret 'bat cave' of this hoverfly.

FACT FILE

ORDER **Diptera** FAMILY **Syrphidae** WING LENGTH **7–12mm** SIMILAR SPECIES **Other hoverflies, bees, wasps and sawflies.**

JAN | FEB | **MAR** | **APR** | **MAY** | **JUN** | **JUL** | **AUG** | **SEP** | **OCT** | **NOV** | DEC

HUMMING HOVERFLY

Syrphus ribesii

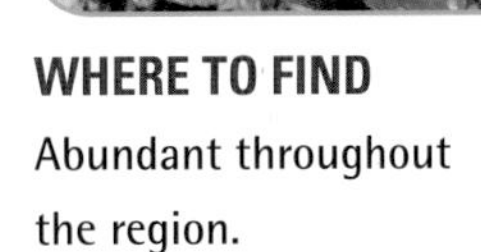

WHERE TO FIND

Abundant throughout the region.

This is the bog-standard hoverfly, looking like a small wasp but easily able to hover motionless in the air on barely visible wings beating at seemingly impossible speed, and darting to and fro. The wings make a hum. The abdomen is bright yellow with thick black cross-bands, which usually have a small peak in the middle and sweep upwards towards the side. The thorax is dull green. The females have yellow legs, something that seems to thrill hoverfly enthusiasts.

EGG (on vegetation) > **LARVA** (often in leaf litter) > **PUPA** > **ADULT**. The larvae are voracious predators of aphids, reportedly eating 50 a day. Hibernates as a larva. Adults are tolerant of freezing temperatures (70 per cent survive at -35 °C).

Have you ever wondered where the gentle summer-day background buzz of the woodland canopy comes from? It is thought to be the communal hum of thousands of male hoverflies vibrating their wings while resting on leaves.

FACT FILE

ORDER **Diptera** FAMILY **Syrphidae** WING LENGTH **7–11.5mm** SIMILAR SPECIES **Many other hoverflies.**

JAN	FEB	MAR	APR	MAY	JUN	JUL	AUG	SEP	OCT	NOV	DEC

BUMBLEBEE CHEILOSIA

Cheilosia illustrata

WHERE TO FIND

Common throughout the region.

This is a portly and furry bumblebee mimic with a buff tail and distinctive spot halfway along the wing known as a 'wing-cloud'. It is a big fan of the plants known as umbellifers, especially hogweed, and it is one of the species you are sure to find where these plants are growing on a warm summer day.

EGG (in soil) > **LARVA** (tunnels into the roots of hogweed) > **PUPA** > **ADULT**. The larvae feed on stems and roots of plants.

The Bumblebee Cheilosia is not just a visual mimic of a bumblebee; it even imitates the buzzes that bees make when alarmed.

FACT FILE

ORDER Diptera FAMILY Syrphidae WING LENGTH 8.5–10mm SIMILAR SPECIES Other hoverflies and bumblebees.

JAN FEB MAR APR **MAY JUN JUL AUG SEP OCT** NOV DEC

GREAT PIED HOVERFLY

Volucella pellucens

This is a true marvel, a fly that is easy to identify. The broad white band across the abdomen, contrasting with black on either side, is unique; notice the wing-cloud too. This very big hoverfly is often found in gardens, but it is really an insect of woods, where males can be seen hovering at human head height in shafts of sunlight. It often feeds at bramble flowers.

EGG > **LARVA** (in wasp nests) > **PUPA** > **ADULT**. The larvae live in the nests of wasps, with several sharing the same nest, where they are scavengers on debris at the bottom of the nest, not predators. They overwinter as pupae in the soil.

★ The adults eat nothing but pollen and nectar, but the larvae eat dead meat (wasp casualties).

WHERE TO FIND

Common to abundant throughout most of the region.

FACT FILE

ORDER Diptera FAMILY Syrphidae WING LENGTH 10–15.5mm SIMILAR SPECIES None.

JAN FEB **MAR APR MAY JUN JUL AUG SEP OCT** NOV DEC

FURRY DRONEFLY

Eristalis intricarius

At first glance this has to be a bumblebee – but it is not, and that is the intention. The Furry Dronefly is a professional mimic of a bumblebee, using the resemblance to protect itself from predators, the colours of the bee being a warning of unpalatability. However, with a second look, the huge eyes, single pair of wings and hovering flight all betray its identity as a hoverfly. The female, with its white abdomen tip, is much the better mimic; the male's 'tail' is brownish. The base of the abdomen, the scutellum, is always yellow, but both sexes are otherwise variable.

EGG (in vegetation) > **LARVA** (in wet, decaying vegetation) > **PUPA** > **ADULT**.

The larvae of this hoverfly like nothing more than slurry pits, cow-dung and other insalubrious dwellings.

WHERE TO FIND

Widespread and common throughout the region, including the north, especially in damper gardens with willows and blackthorn.

FACT FILE

ORDER Diptera FAMILY Syrphidae WING LENGTH 8–12mm SIMILAR SPECIES Mainly bumblebees, but these have much smaller eyes.

JAN | FEB | MAR | APR | MAY | JUN | JUL | AUG | SEP | OCT | NOV | DEC

COMMON DRONEFLY

Eristalis tenax

This is a fly that goes around mimicking a Honey Bee (p. 152), using the latter's notorious sting and warning colouration as a cover for its own palatable self. When flying, it even holds its hindlegs down in an attempt to look as though it has a pollen sac. However, the two wings and complete lack of a 'waist' between the thorax and abdomen easily distinguish it. It is a heavily built fly with variable amounts of black on its abdomen; the black absorbs heat. There is also usually a pale yellowish band across the abdomen. It flies even in the middle of winter and feeds on numerous flowers, including those of ivy. Males are very territorial and fight off any insects in their patch.

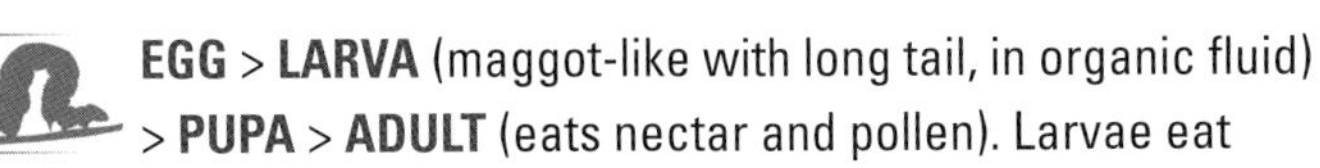

EGG > **LARVA** (maggot-like with long tail, in organic fluid) > **PUPA** > **ADULT** (eats nectar and pollen). Larvae eat bacteria in enriched organic goo, even in silage and sewage. Overwinters as an adult.

This is a super-fly. Remove the wings and it keeps feeding as if nothing has happened. Remove the head and it can live for three days.

WHERE TO FIND

Very common everywhere in the region – and indeed, worldwide.

FACT FILE

ORDER Diptera FAMILY Syrphidae WINGSPAN 10–13mm SIMILAR SPECIES Honey Bee. A good distinction is that the hoverfly's eyes meet at the top of the head.

JAN FEB MAR APR **MAY JUN JUL AUG SEP** OCT NOV DEC

THICK-HEADED FLY

Sicus ferrugineus

WHERE TO FIND

Common in most of the region north to southern Scandinavia and the Baltic States. Widespread in Britain and Ireland.

At first sight – assuming you do not simply overlook it as just another fly basking on a sunlit summer flower – you might notice that this is a rather deliciously coloured insect, with a rich dark honey-coloured body and bright yellow front. The proportions are odd, however, with the insect having a curious hunched back, narrow abdomen that curves downwards, eponymous bloated head that is as wide as the thorax, and curious short proboscis. Its lifestyle is even odder, given away by its other name of 'Bee Grabber', for the fly hugs a bee in flight and lays an egg into its abdomen. Thereafter, the larva develops inside, gradually consuming the host and only pupating when the unfortunate creature dies.

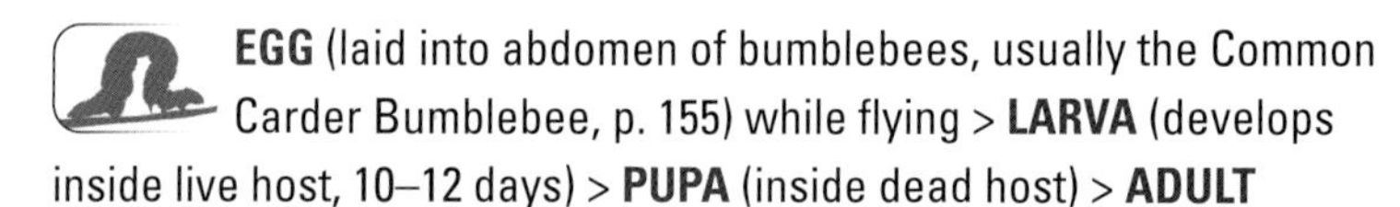

EGG (laid into abdomen of bumblebees, usually the Common Carder Bumblebee, p. 155) while flying > **LARVA** (develops inside live host, 10–12 days) > **PUPA** (inside dead host) > **ADULT** (eats nectar and pollen from thistles, bramble and many garden flowers). Overwinters as a pupa and emerges in early summer.

This fly frequently competes with a related parasite of the same family, with eggs of both species often being laid in the same unfortunate individual host. The larva of this species usually develops more quickly and has a slight advantage.

FACT FILE

ORDER Diptera FAMILY Conopidae BODY LENGTH 8–13mm SIMILAR SPECIES The rich colour and obvious yellow head are quite distinctive. Note the two wings, not four as in bees and wasps.

JAN FEB MAR APR **MAY JUN JUL AUG** SEP OCT NOV DEC

THISTLE GALL FLY

Urophora cardui

This is a tiny fly that would be anonymous but for its beautiful, boldly marked wings that give rise to the generic name of 'picture-winged fly', of which there are many species. This one has a distinctive thick 'M'-mark on each wing and a white spot on the base of the abdomen. It also has a claim to fame. The larvae burrow into the stem of a thistle and produce a swelling known as a 'gall'. This species only attacks creeping thistles, an abundant species of waste places and gardens, and the bulbous galls are up to 10cm long and very obvious, especially when a thistle dies back in the autumn. There are chambers for several larvae and the adults emerge only when a gall starts to rot, which is in midsummer.

WHERE TO FIND

Occurs throughout the region.

EGG (batch at tip of young shoots of thistle) > **LARVA** (grubs, burrow into stem and make gall, three instars) > **PUPA** (in gall, early spring) > **ADULT**. Adults feed on pollen, nectar and plant debris. Overwinters as a larva.

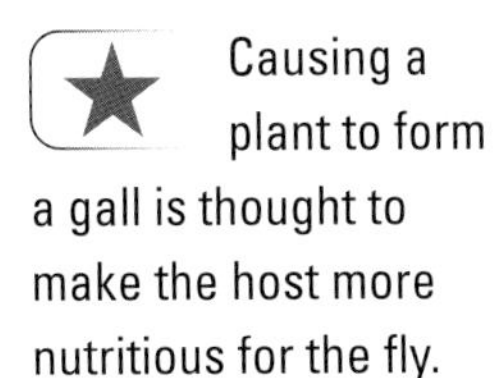

Causing a plant to form a gall is thought to make the host more nutritious for the fly.

FACT FILE

ORDER Diptera FAMILY Tephritidae BODY LENGTH 6–8mm SIMILAR SPECIES There are many other species of similar pattern and size, which use different plant hosts.

JAN FEB MAR APR **MAY** **JUN** **JUL** **AUG** **SEP** OCT NOV DEC

SNAIL-KILLING FLY

Coremacera marginata

WHERE TO FIND
Most common in the south of the region (to the southern UK), but widely – if locally – distributed.

This is a small, overlooked fly, but its life cycle is a fine example of the dramatic, impersonal cruelties that are part of the day-to-day life of some of the garden's minibeasts. The adults are good-looking, with their net-curtain wings and mauve-streaked eyes, and all they do is drink dew and nectar. The grubs, however, feed themselves by burying their way into the flesh of land-living snails and consuming them from within.

EGG > **LARVA** (grub, each individual feeds on 1–3 different terrestrial snails) > **PUPA** > **ADULT**.

★ The vital organs of a snail go last, suggesting that the host can feel its demise.

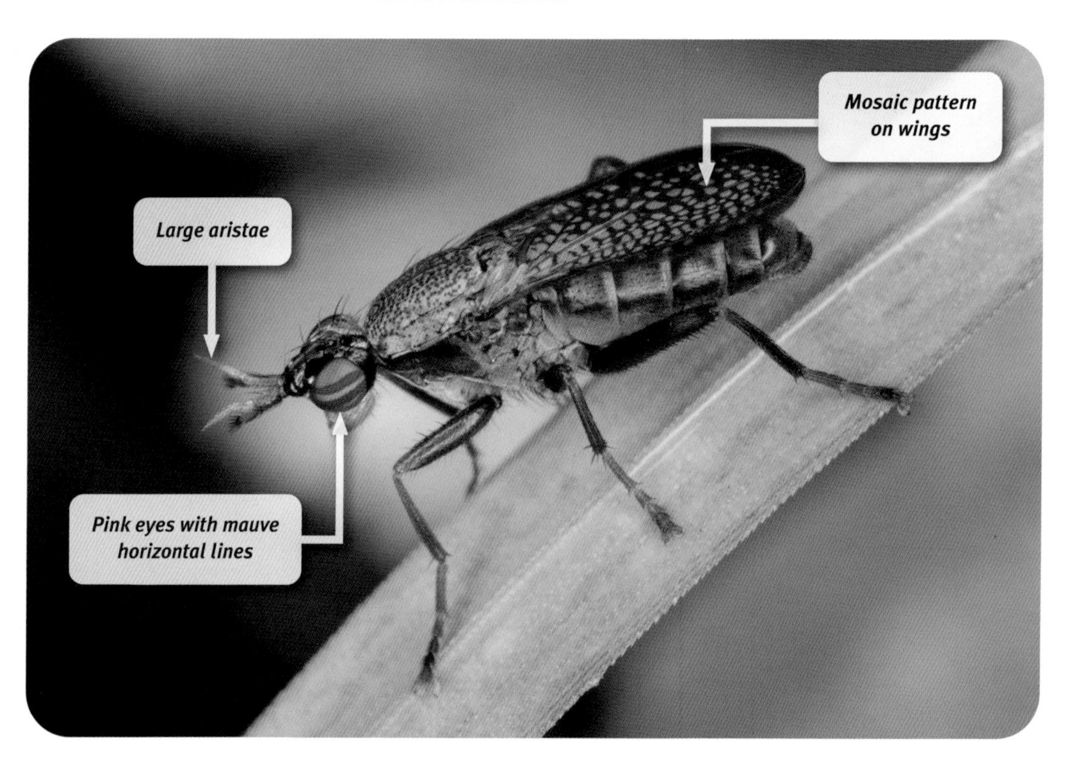

FACT FILE

ORDER Diptera FAMILY Sciomyzidae BODY LENGTH 7–10mm SIMILAR SPECIES The beautiful lacy wings of this species set it apart from most other flies.

JAN FEB **MAR APR MAY JUN JUL AUG SEP** OCT NOV DEC

YELLOW DUNG FLY

Scathophaga stercoraria

This species is glamorous for a fly, with a battery of brightly coloured hairs giving the body a golden, almost orange glow; males are more brightly coloured than females. The fly's life is less glamorous, since it spends much time on the dung of animals such as cows and horses. It also commonly visits flowers and gardens, and may be attracted to artificial light.

EGG (laid on dung) > **LARVA** (in dung) > **PUPA** (buried in dung for 20 days > **ADULT**.

Girl meets boy on dung; it does not always go well, since a visiting female may be smothered and badly injured by the scrum of amorous males. Some individuals of this species are known to suffer from sexually transmitted diseases.

WHERE TO FIND

Occurs throughout the region where livestock is kept.

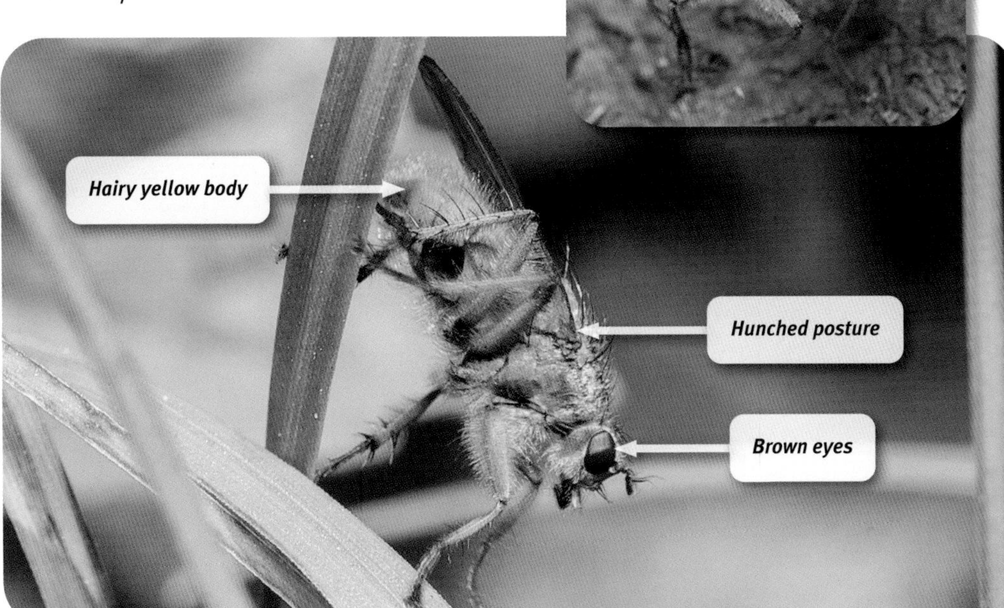

ABOVE: *LATERAL VIEW*; TOP: *ON DUNG*

FACT FILE

ORDER Diptera FAMILY Scathophagidae BODY LENGTH 8–10mm SIMILAR SPECIES Some other flies have a similar colour, but the dung habitat is diagnostic.

JAN FEB MAR APR **MAY JUN JUL AUG SEP OCT** NOV DEC

KITE-TAILED ROBBERFLY

Tolmerus atricapillus

WHERE TO FIND
Widespread and common throughout the region.

Robberflies are big, bristly legged, long-bodied flies that just look like trouble – and they are, at least to other insects. They perch in the sun waiting for something to fly past (sometimes a hoverfly), then take off to ambush it in mid-air. The spidery legs hold the prey under the body and robberflies have a horny proboscis to puncture the exoskeleton and drink the fluids; meanwhile, the head is protected from the struggles of the prey by bristles on the face. Despite their appearance, robberflies do not trouble people, although they are alleged to bite occasionally.

EGG (laid into buds of flowers) > **LARVA** (in soil) > **PUPA** (in soil) > **ADULT**. The larvae are predatory. Hibernates as a larva, for two years.

The toxins released from the mouthparts to paralyze prey are the most potent known from any fly.

FACT FILE

ORDER Diptera FAMILY Asilidae BODY LENGTH 12–15mm SIMILAR SPECIES Horseflies are similar in size.

JAN | FEB | MAR | APR | MAY | JUN | JUL | AUG | SEP | OCT | NOV | DEC

HOUSE FLIES

Muscidae spp.

WHERE TO FIND

Abundant everywhere.

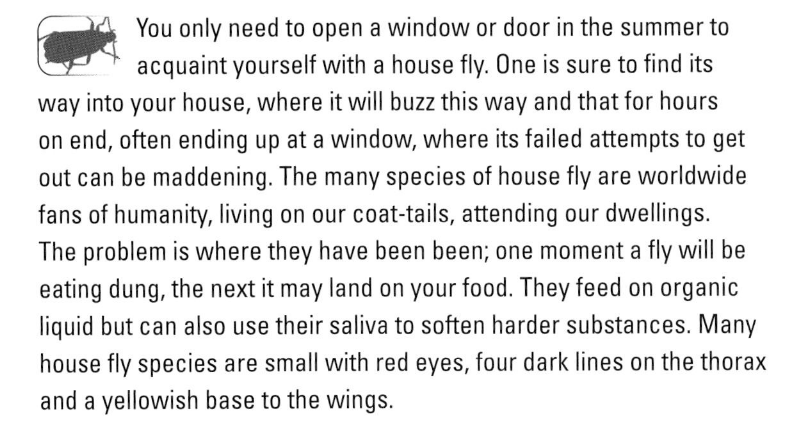

You only need to open a window or door in the summer to acquaint yourself with a house fly. One is sure to find its way into your house, where it will buzz this way and that for hours on end, often ending up at a window, where its failed attempts to get out can be maddening. The many species of house fly are worldwide fans of humanity, living on our coat-tails, attending our dwellings. The problem is where they have been been; one moment a fly will be eating dung, the next it may land on your food. They feed on organic liquid but can also use their saliva to soften harder substances. Many house fly species are small with red eyes, four dark lines on the thorax and a yellowish base to the wings.

EGG Those of Common House Fly *Musca domestica* laid on organic waste, carrion or faeces; up to 500 in batches, hatch within a day > **LARVA** (maggot, in bodies or faeces, 14–30 days) > **PUPA** (in soil, 2–20 days) > **ADULT**. Adults feed on organic matter of all kinds.

Despite its size, a Common House Fly can travel for 16km or more. Males initiate courtship by basically flying towards females and bumping into them. A well-fed fly defecates every 4.5 minutes.

GRAPHOMYA MACULATA

FACT FILE

ORDER Diptera FAMILY Muscidae BODY LENGTH 7–8mm SIMILAR SPECIES Cluster Fly *Pollenia rudis* – but the Common House Fly has a more orange-tinged abdomen, whereas the Cluster Fly is more silver and black. There are many species of house fly and other similar flies.

JAN | FEB | MAR | APR | MAY | JUN | JUL | AUG | SEP | OCT | NOV | DEC

BLUEBOTTLE

Calliphora vomitoria

WHERE TO FIND
Very common everywhere.

Not one of the garden's secrets – you have probably swatted this species or sworn at it many times. It is a large, annoying, buzzy fly. It has none of the compensating glamour of a Greenbottle (opposite), being much larger, hairier and an unexciting shade of dark blue. Not even fetching orange cheeks below the eyes help. The Bluebottle is undoubtedly useful, playing a vital part in cleaning up dead bodies; the maggots eat flesh and, although they look revolting in insalubrious places in the house, they are harmless. The adults, on the other hand, can spread infections by landing on our food with their dirty feet.

EGG (laid in batches, up to 200, on carrion) > **LARVA** (inside carcasses of all kinds, or faeces) > **PUPA** (in soil) > **ADULT**. Life cycle takes a few weeks, depending on the temperature. Adults feed on nectar.

Bluebottles that detect a corpse release a chemical signal that attracts other flies. Having many larvae feeding at once keeps it nice and warm. Flies release blobs of fluid from their feet, which helps them cling to surfaces. The scientific name '*Calliphora*' means 'bearer of beauty'!

FACT FILE

ORDER Diptera FAMILY Calliphoridae BODY LENGTH 10–14mm SIMILAR SPECIES There are several similar species.

JAN | FEB | MAR | APR | MAY | JUN | JUL | AUG | SEP | OCT | NOV | DEC

GREENBOTTLE

Lucilia sericata

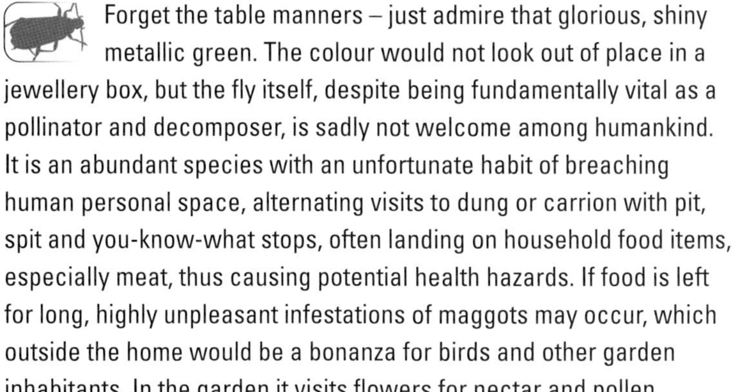

Forget the table manners – just admire that glorious, shiny metallic green. The colour would not look out of place in a jewellery box, but the fly itself, despite being fundamentally vital as a pollinator and decomposer, is sadly not welcome among humankind. It is an abundant species with an unfortunate habit of breaching human personal space, alternating visits to dung or carrion with pit, spit and you-know-what stops, often landing on household food items, especially meat, thus causing potential health hazards. If food is left for long, highly unpleasant infestations of maggots may occur, which outside the home would be a bonanza for birds and other garden inhabitants. In the garden it visits flowers for nectar and pollen.

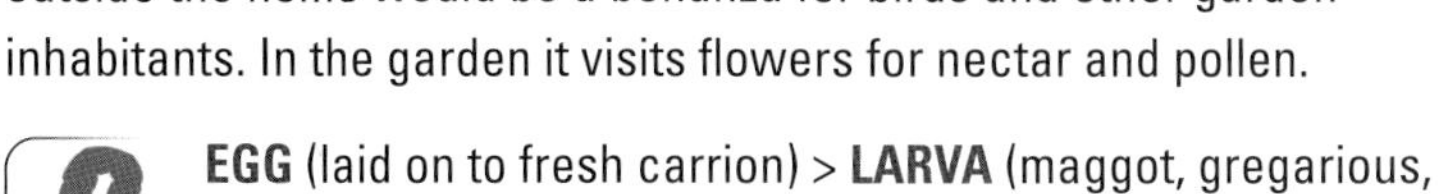

EGG (laid on to fresh carrion) > **LARVA** (maggot, gregarious, in rotting flesh) > **PUPA** (buried in soil) > **ADULT**.

The appearance and development of Greenbottle fly maggots is of critical importance in forensic investigations. Many a murderer has been caught out by blowflies. Males court by tapping a female's body with their foreleg. Some use the left leg and some the right. Doctors sometimes use 'maggot tea bags' to treat wounds.

WHERE TO FIND

Common and often abundant throughout the region.

FACT FILE

ORDER **Diptera** FAMILY **Calliphoridae** BODY LENGTH **8–10mm** SIMILAR SPECIES **There are several similar species.**

JAN | FEB | MAR | APR | MAY | JUN | JUL | AUG | SEP | OCT | NOV | DEC

FLESH FLIES

Sarcophagidae spp.

WHERE TO FIND
Common and often abundant throughout the region.

A flesh fly is a handsome beast, with its big red eyes, glinting chequerboard pattern on the abdomen and long, hairy legs that splay out at rest. It is a big, powerful insect with fast flight. It is easy to tell males from females, because in males the eyes are almost touching, while the females' eyes are well separated. This species abounds in gardens but has less of a habit of finding its way into houses than other common flies. Unusually, it gives birth to live maggoty young, straight on to carrion.

EGG (internal) > **LARVA** (gives birth to live maggots, in decaying flesh) > **PUPA** (buried in soil) > **ADULT**. Overwinters as a pupa.

The remarkable trick of giving birth to live maggots helps the latter to feed on rapidly decomposing flesh, which never lasts long.

FACT FILE

ORDER Diptera FAMILY Sarcophagidae BODY LENGTH 10–15mm SIMILAR SPECIES There are dozens of similar species, but flesh flies as a whole are quite distinctive.

JAN	FEB	MAR	APR	**MAY**	**JUN**	**JUL**	**AUG**	SEP	OCT	NOV	DEC

SCORPIONFLY

Panorpa communis

The aliens have invaded the Earth, but we have not noticed, because they fly sluggishly in low vegetation such as brambles on warm, sunny days. A close look reveals all, though: the strange, scary-looking mouthparts (rostrum) that point downwards, and especially the male's enlarged tail (the genitals), curled over the back, which looks like the sting of a scorpion – if cornered the animal will use it to pretend to sting. The wings are elongated and spotted, and the body is long and slender, too. No other insect looks quite like this and there are no close relatives. The scorpionflies belong to an ancient order of insects. They are not true flies, but they are distantly related to them.

EGG (in moisture) > **LARVA** (caterpillar-like, feeds on dead animals) > **PUPA** (in soil) > **ADULT**. Adults scavenge and also eat fruits. Overwinters as a pupa.

★ Scorpionflies often steal insect food from spiders' webs, a pretty risky meal. They sometimes feed on bird droppings.

WHERE TO FIND

Common throughout the region.

FACT FILE

ORDER Mecoptera FAMILY Panorpidae BODY LENGTH 10–15mm SIMILAR SPECIES Mayflies, caddisflies, lacewings and large flies may all look similar, but the Scorpionfly's down-pointing mouthparts are distinctive.

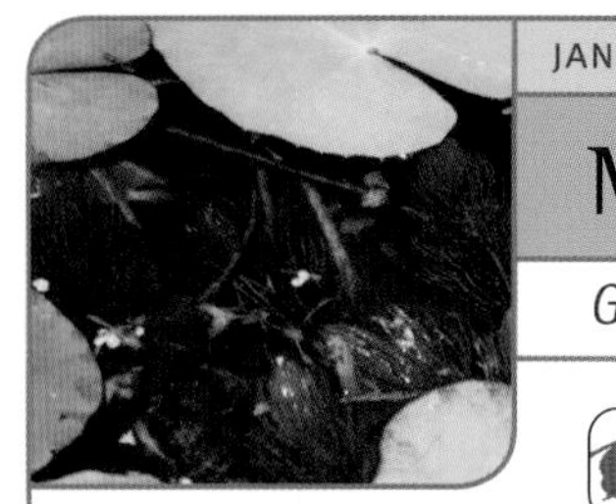

JAN | FEB | MAR | APR | **MAY** | **JUN** | **JUL** | **AUG** | **SEP** | **OCT** | NOV | DEC

MOTTLED SEDGE CADDISFLY

Glyphotaelius pellucidus

WHERE TO FIND

This species is common and widespread throughout the region. Many other caddisflies occur in all areas.

Caddisflies count among the few garden insects best known for their larvae. They are aquatic, and pond-dippers frequently find them inside their remarkable cases, which envelop them up to their legs and head. The extraordinary case of this species is made of large, round, flat pieces of leaf stuck together with the animal's body slime. The larvae move about and their cases grow as they do. The adults are nocturnal and they look like moths, with the wings held over the body like a roof, but they have small hairs on the wings and lack a proboscis. The antennae are often as long as the body. Watch over the garden pond at dusk for these insects darting about; they are also attracted to light, including moth traps.

EGG (cast over water) > **LARVA** (aquatic, within case, six legs) > **PUPA** (in case) > **ADULT**. Larvae feed on organic matter. Adults do not feed but may sip nectar or water. Overwinters as a larva.

Caddisflies look like weak fliers, but they can travel as much as 5km, so can easily find new garden ponds. The larvae breathe through gills on the abdomen.

FACT FILE

ORDER Trichoptera FAMILY Limnephilidae FOREWING LENGTH 12–17mm SIMILAR SPECIES There are many species of caddisfly, but this is the only one with a notch on the outer margin of the forewing. Moths, mayflies and alderflies fly in similar fashion.

JAN FEB MAR APR MAY **JUN JUL AUG** SEP OCT NOV DEC

SMALL SKIPPER

Thymelicus sylvestris

Skittering over grass and flowers on golden whirring wings, the Small Skipper breaks down our idea of what a butterfly is. Colourful and day flying, it nevertheless has hooked antennae, a moth feature, as opposed to clubbed ones. Moreover, when it lands, on a thistle-head for instance, it holds up its forewings and keeps its hindwings flat, which is unusual for a butterfly and characteristic. This is a warm-weather insect of grassy places, which commonly finds its way into gardens and feeds on thistles and clovers, among other plants. It usually flies low down and weakly.

EGG (on leaf sheaths of plants, batches of 2–15) > **LARVA** (caterpillar, eats egg case and hibernates; feeds in spring) > **PUPA** (tents of leaves spun with silk) > **ADULT**. The larval food plant is an abundant grass called Yorkshire Fog. The eggs hatch in August; overwinters as a larva.

It needs a certain length of grass to survive, simply because the caterpillars need long blades from which to roll protective tubes around themselves.

WHERE TO FIND

Common, but absent from Scotland, Ireland, Norway, Sweden and Finland.

FACT FILE

ORDER Lepidoptera FAMILY Hesperiidae WINGSPAN 27–34mm SIMILAR SPECIES There are several similar skipper species. The Essex Skipper *T. lineola* is almost identical, but has black, not brown, bases to the tips of its antennae. The Large Skipper *Ochlodes venatus* is larger, and has brown edges to the wings. Both reach southern Norway and Sweden.

JAN FEB MAR APR MAY JUN JUL AUG SEP OCT NOV DEC

BRIMSTONE

Gonepteryx rhamni

WHERE TO FIND
Very common everywhere, except Scotland and northern Scandinavia.

This much-loved spring butterfly emerges with the daffodils and is of a similar fresh colour and breeziness. It hibernates as an adult and can occasionally be seen as early as January on warm days. Many people do not realize that it has a second career as a summer butterfly, with the early adults laying eggs that produce a new generation of adults in June. The species is large, with rapid, powerful wingbeats. The male is an intense sulphur-yellow, the female greener. The butterfly invariably closes its wings at rest, and quite suddenly morphs into a leaf shape. A close-up view reveals that the face has psychedelic pink markings.

LATERAL VIEW OF FEMALE

EGG (laid singly) > **LARVA** (caterpillar) > **PUPA** (low down among vegetation) > **ADULT**. The larval food plant is a shrub, buckthorn. The eggs hatch in midsummer. It overwinters as an adult, which may live for 11 months.

It is thought that the word 'butterfly' comes from the butter-like colour of male Brimstones. The species has an unusually long proboscis, enabling it to feed deep in flowers of shrubs such as buddleia.

FACT FILE

ORDER **Lepidoptera** FAMILY **Pieridae** WINGSPAN **60–74mm** SIMILAR SPECIES **The whites (pp. 103–104) have yellowish undersides to the wings, but are never as bright. The whites' wings are a different shape, more rounded at the tip, in contrast to the sharp points of the Brimstone's wings.**

JAN	FEB	MAR	**APR**	**MAY**	**JUN**	**JUL**	**AUG**	**SEP**	**OCT**	**NOV**	DEC

LARGE WHITE

Pieris brassicae

WHERE TO FIND
Abundant everywhere.

This species lives up to its name as a big white butterfly and is abundant everywhere throughout the summer months. Its nickname of 'Cabbage White' betrays its caterpillar's fondness for the leaves of cabbage and other brassicas (the Small White, p. 104, shares the same nickname and habits). A big white butterfly with thick bold black wing-tips is always a Large White; at rest it also always shows big black spots on the underside. However, there are differences between the male and female. The latter is larger and has big, inky-black spots on the upperside as well as the underside, which are lacking in the male. Males can also look quite small, so take care not to mistake them for Small Whites.

EGG (bunches of 30 plus) > **LARVA** (caterpillar) > **PUPA** > **ADULT**. Larval food plants are cabbages and their relatives. Overwinters as a pupa. In the late summer many arrive as immigrants from further south.

Most butterfly species take great care in finding a suitable egg-laying site, but female Large Whites home in quickly using blue-green colour, or using polarized light.

ABOVE: *FEMALE*; TOP: *MALE*

FACT FILE

ORDER Lepidoptera FAMILY Pieridae WINGSPAN 53–70mm SIMILAR SPECIES The Small White is smaller, not surprisingly. The Brimstone (opposite) is yellow or green, has different-shaped wings and looks like a leaf at rest.

JAN | FEB | MAR | **APR** | **MAY** | **JUN** | **JUL** | **AUG** | **SEP** | **OCT** | NOV | DEC

SMALL WHITE

Pieris rapae

WHERE TO FIND

Abundant throughout the region.

The small version of the 'Cabbage White', this butterfly is remarkably similar in habits to the Large White (p. 103), using the same food plants and having a very similar appearance. The best distinction, apart from size, is that the black tips to the forewings, and the mid-wing spots, have a faded look so that – had they been drawn with a pencil – they would have been half rubbed out. The spots are also smaller and less obvious. The male Small White's markings are even less clear-cut than the female's, but they have a central spot that the male Large White lacks.

EGG (laid singly on food plant) > **LARVA** (caterpillar, green for camouflage, lacks noxious body fluids) > **PUPA** > **ADULT**. Larval food plants are cabbages and their relatives. Overwinters as a pupa.

Small Whites use ultraviolet cues to find good foraging flowers.

RIGHT: *EGG*; BELOW LEFT & RIGHT: *ADULT*

FACT FILE

ORDER Lepidoptera FAMILY Pieridae WINGSPAN 38–57mm SIMILAR SPECIES Green-veined White *P. napi*. This is another common garden butterfly, of similar size to the Small White, and with distinctive green veins on the underwings; wingspan 40–52mm; April–September.

JAN FEB MAR **APR MAY JUN JUL AUG** SEP OCT NOV DEC

ORANGE-TIP

Anthocharis cardamines

As much a sign of spring as the first swallow, the male of this delicate butterfly looks as though it has dipped its wings into condensed orange juice, as it flies around the new leaves and fresh blooms with a detached, fluttery haste. The female lacks the orange, instead having black wing-tips and central comma-like spots, and it can be written off as yet another Small White (opposite). However, when at rest, with wings closed, both sexes show off a gloriously intricate latticework of mottled green and white on the hindwing, like an expensive artisan design.

WHERE TO FIND

Fairly common throughout the region.

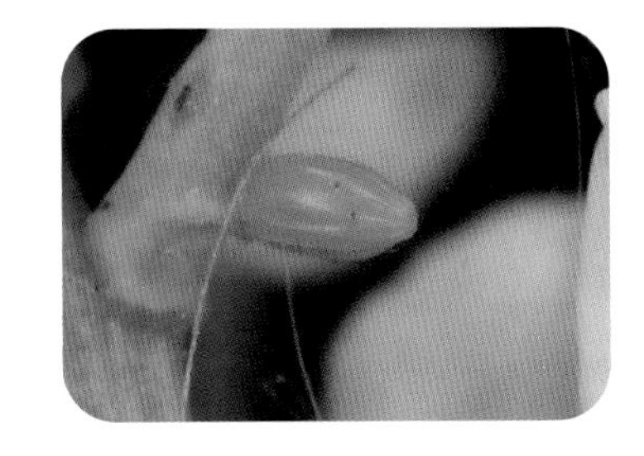

EGG (laid singly at base of flowerhead) > **LARVA** (caterpillar) > **PUPA** > **ADULT**. The main larval food plant is cuckoo-flower and its relatives. Overwinters as a pupa.

If several eggs are laid on one plant, the caterpillars will cannibalize their rivals' eggs. Egg-laying females leave behind a chemical message that the site has been taken.

TOP RIGHT: *EGG*; MIDDLE RIGHT: *FEMALE UNDERSIDE*
ABOVE LEFT: *FEMALE*; ABOVE RIGHT: *MALE*

FACT FILE

ORDER **Lepidoptera** FAMILY **Pieridae** WINGSPAN **40–52mm** SIMILAR SPECIES **Whites, especially the Small White.**

JAN	FEB	MAR	APR	MAY	JUN	JUL	AUG	SEP	OCT	NOV	DEC

SMALL TORTOISESHELL

Aglais urticae

WHERE TO FIND

Very common throughout the region, in almost any habitat.

At the midsummer scrum at a buddleia, it is surprising how this brilliantly coloured butterfly can be overlooked, especially next to the 'look-at-me' patterns of the Peacock (p. 108) and Red Admiral (opposite). Nevertheless, it has a stunning colour scheme all of its own, with a unique necklace of blue 'beads' running down the outside edges of both wings. The equally distinctive tortoiseshell pattern along the front edges of the forewings is also unique. Much smaller than its near relatives, the very common Small Tortoiseshell looks more triangular at rest and has a more direct flight on faster wingbeats, often flying low down. When it folds its wings shut, the dark inside, pale outside pattern is also diagnostic.

EGG (on underside of leaf in sunlight) > **LARVA** (caterpillar) > **PUPA** (suspended from vegetation) > **ADULT**. The main larval food plants are nettles; the caterpillars are sociable and live in webs. They are greener than similar species. One of our hibernating butterflies, it overwinters as an adult. (Eggs May, then July; larva May–June, July–August; pupa June and mid-August)

UPPERSIDE

★ The length of the day of hatching determines whether the larvae will develop into breeding adults or hibernating adults.

FACT FILE

ORDER Lepidoptera FAMILY Nymphalidae WINGSPAN 45–62mm SIMILAR SPECIES The Red Admiral, Painted Lady and Comma (pp. 107, 109 and 110) are superficially similar, but all are larger.

JAN | FEB | **MAR** | **APR** | **MAY** | **JUN** | **JUL** | **AUG** | **SEP** | **OCT** | **NOV** | DEC

RED ADMIRAL

Vanessa atalanta

This extrovert big, bold and colourful butterfly rivals the Peacock (p. 108) in size and abundance, and the two can be confused. However, the Red Admiral entirely lacks the Peacock's 'eyes', instead having black-and-white wing-tips. There is a broad band of tangerine-orange across the wing and another on the trailing edge. With wings closed the colour of the upperwing is mirrored on the underside. A supremely adaptable butterfly, the Red Admiral sips nectar from flowers and is attracted to rotting fruits, blackberries and ivy blossom. Up until recently the early stages could not survive northern winters, so the entire population every year would be made up of immigrants rather than butterflies that had bred locally. Even now it is most common in autumn.

WHERE TO FIND

Common throughout the region. British and Scandinavian populations depend on annual immigration.

EGG (laid singly on upperside of leaf) > **LARVA** (caterpillar) > **PUPA** > **ADULT**. The larval food plant is nettle, but unlike its relatives the mainly black caterpillar is solitary. One of our hibernating butterflies, it overwinters as an adult.

A butterfly chrysalis looks immobile, but that of the Red Admiral (and others) can shake vigorously when disturbed. This seems to be a good defence against parasitic wasps.

UPPERSIDE

FACT FILE

ORDER Lepidoptera FAMILY Nymphalidae WINGSPAN 64–78mm SIMILAR SPECIES Small Tortoiseshell, Peacock and Painted Lady (pp. 106, 108 and 109).

JAN | FEB | **MAR** | **APR** | **MAY** | **JUN** | **JUL** | **AUG** | **SEP** | OCT | NOV | DEC

PEACOCK

Aglais io

WHERE TO FIND

Very common almost everywhere, but only found in the south of Norway, Sweden and Finland.

This is a big, powerful and spectacular butterfly with the sort of colour scheme that you might expect from a tropical species. The Peacock is common, widespread and has a long flight season. It is instantly recognizable by the prominent 'eyes' at the corners of both forewings and hindwings, like those of the eponymous peacock, the bird. These are set against a marvellous intense fiery red-orange background. This bling all instantly disappears when the butterfly lands, and it transforms into an impressively camouflaged dead-leaf pattern that is easily overlooked. When flying any distance, it combines series of flaps with glides, and may look like a leaf caught in the wind.

EGG (large clusters on undersides of leaves) > **LARVA** (caterpillar) > **PUPA** > **ADULT**. The larval food plant is nettle. The black, spiny caterpillars live communally inside an obvious 'web'. Overwinters as an adult. The eggs usually appear in May, the caterpillars in June, and the chrysalis in July.

If disturbed during hibernation, the Peacock may rub its wings together to make a quiet hissing sound. Males feed in the morning, then form mating territories in the afternoon.

BELOW LEFT: *UPPERSIDE*; BELOW RIGHT: *UNDERSIDE*

FACT FILE

ORDER Lepidoptera FAMILY Nymphalidae WINGSPAN 63–75mm SIMILAR SPECIES With a proper look, it is easily distinguished from the Red Admiral (p. 107) and Small Tortoiseshell (p. 106).

JAN FEB **MAR APR MAY JUN JUL AUG SEP OCT** NOV DEC

PAINTED LADY

Vanessa cardui

Famous for its fast, powerful flight and remarkable wandering abilities, the Painted Lady is one of our most charismatic butterflies. Similar in size to a Red Admiral (p. 107), and with a similar black-and-white wing-tip, this butterfly differs by having an overall orange or even salmon-pink hue that is very much its own. The same colour shows on the underside when the butterfly is at rest. It can be seen anywhere, in any garden, and feeds at the nectar of multiple plants. Thistles are favoured in the wild. Spring individuals arrive directly from North Africa, like migrating birds.

EGG (laid singly) > **LARVA** (caterpillar) > **PUPA** > **ADULT**. The larval food plants are thistles. The black, spiny caterpillars live singly inside a tent. It does not survive the winter in Britain. Eggs laid in May–July may produce breeding adults; those laid in the late summer and autumn never survive.

A great migrant, the Painted Lady is sometimes seen flying south in autumn, for example, away from the British coastline.

WHERE TO FIND

Occurs throughout the region. Numbers vary enormously from year to year, so sometimes very common, at other times scarce.

LEFT: *UPPERSIDE*; BELOW: *UNDERSIDE*

FACT FILE

ORDER Lepidoptera FAMILY Nymphalidae WINGSPAN 58–74mm SIMILAR SPECIES It is the size of a Red Admiral, but the orange hue is distinctive.

JAN	FEB	**MAR**	**APR**	**MAY**	**JUN**	**JUL**	**AUG**	**SEP**	OCT	NOV	DEC

COMMA

Polygonia c-album

WHERE TO FIND

Widespread and common, except in the far north of Scandinavia.

This is a truly unmistakable butterfly, with unique ragged edges to its wings. Medium sized, it flies with surprising power, alternating rapid wingbeats and glides, making it look larger. When the wings are closed it closely resembles a crinkled dead leaf, but always shows a white 'U'-shaped mark on the hindwing, the eponymous comma. Common in gardens, the butterfly visits blooms of plants such as buddleia and is also attracted to rotting fruits. It is one of the earliest butterflies to appear in spring.

EGG (laid singly) > **LARVA** (caterpillar, initially spins a web) > **PUPA** (suspended in thick vegetation) > **ADULT**. The larval food plants include nettles, Hops, elm and willow. The caterpillar resembles a bird dropping. Overwinters as an adult.

Once rare and declining in Britain, the Comma's fortunes were approaching a full stop, but then it suddenly 'decided' that nettle was an acceptable food plant and has not looked back since.

LEFT: *UPPERSIDE*; RIGHT: *UNDERSIDE – SHOWING 'COMMA'*

FACT FILE

ORDER Lepidoptera FAMILY Nymphalidae WINGSPAN 50–64mm SIMILAR SPECIES Its powerful flight may recall a larger butterfly such as the Painted Lady (p. 109), while it is of a similar size to the Small Tortoiseshell (p. 106).

JAN FEB **MAR APR MAY JUN JUL AUG SEP OCT** NOV DEC

SPECKLED WOOD

Parage aegeria

The Speckled Wood looks like a monochrome version of a more colourful butterfly, although it is still a striking and assertive one. It could be called the dappled shade butterfly, not only because of its wing markings, but also because the males compete over shafts of sunlight in woods and bushy areas, spiralling upwards in a levitating boxing match between lightweights. The Speckled Wood commonly visits shady places in gardens, often perching on leaves rather than flowers.

EGG (singly on grass leaves) > **LARVA** (caterpillar, green, and camouflaged on edge of grass leaf) > **PUPA** > **ADULT**. The larval food plants are grasses. Overwinters as a larva or pupa.

★ In spring and autumn the female lays eggs in warm spots, but in summer prefers heavily shaded locations.

WHERE TO FIND

Common although often patchily distributed throughout the region, north to central Scandinavia.

UPPERSIDE

FACT FILE

ORDER **Lepidoptera** FAMILY **Nymphalidae** WINGSPAN **46–56mm** SIMILAR SPECIES **Distinctive.**

JAN | FEB | MAR | APR | MAY | **JUN** | **JUL** | **AUG** | **SEP** | OCT | NOV | DEC

MEADOW BROWN

Maniola jurtina

WHERE TO FIND

Very common throughout the region, north to southern Norway, Sweden and Finland.

This definitely is not the most glamorous of our butterflies, but it is very common, reaches right into urban areas and even flies in chilly, overcast conditions, so we should be grateful for it. It is one of several 'brown' butterflies that might occur in the garden. The male is dark brown and has a single black eye-spot near the tip of each wing, with a small bright white spot in the centre and an orange corona around. The female shares this feature, but has a modest orange-yellow flash on the forewing. When at rest, both show a single eye-spot inside an orange streak.

EGG (laid singly on grass) > **LARVA** (caterpillar, feeds on grasses) > **PUPA** (at base of grass or in litter) > **ADULT**. The larval food plants are grasses. Overwinters as a larva.

Adult Meadow Browns live for just 3–12 days.

LEFT: *UPPERSIDE*; RIGHT: *UNDERSIDE*

FACT FILE

ORDER Lepidoptera FAMILY Nymphalidae WINGSPAN 40–60mm SIMILAR SPECIES The Gatekeeper *Pyronia tithonus* (southern Britain and near Continent) is smaller than the Meadow Brown, with two white-centred black spots above and below. The wings are orange with a neat brown edge. The Ringlet *Aphantopus hyperanthus* (throughout the north to central Scandinavia, and uncommon in northern France) is unyieldingly dark above, but when at rest shows a series of yellow-edged eye-spots that seem to well up from the body like bubbles.

JAN FEB MAR **APR MAY JUN JUL AUG SEP OCT** NOV DEC

SMALL COPPER

Lycaena phlaeus

The Small Copper is a fiery sprite with orange wings and a hot temper. It is small and fast, and easy to lose in flight. It spends much time sunning itself low down, typically on the ground. The shining, coppery-orange forewings shared by both sexes, together with the neat dark brown spots, are unique. Males defend a patch of ground and will fly up at almost anything that passes, such is their pugnacity. This is mainly a meadow and rough ground butterfly, but it often wanders into gardens.

EGG (under leaves) > **LARVA** (caterpillar) > **PUPA** (tended by ants in leaf litter) > **ADULT**. Larval food plants are sorrels. Overwinters as a caterpillar. There can be as many as four generations in a year, with the last adults on the wing in October.

The caterpillar produces fluids rich in nutrients from pores near the head and honey gland, which are licked by ants.

WHERE TO FIND

Common throughout the region.

UPPERSIDE

FACT FILE

ORDER **Lepidoptera** FAMILY **Lycaenidae** WINGSPAN **26–36mm** SIMILAR SPECIES **The Small Skipper (p. 101) is of a similar size, but lacks the black spots.**

JAN FEB MAR APR **MAY JUN JUL AUG SEP OCT** NOV DEC

COMMON BLUE

Polyommatus icarus

WHERE TO FIND
Common throughout the region.

The Common Blue is not as common in gardens as the Holly Blue (opposite), mainly because it is a meadow and roadside butterfly that abounds in dry, warm, flowery places where bird's-foot trefoil, a yellow wildflower, grows. Where it does occur, however, perhaps in a bigger or wilder garden, it is often abundant. The male is a hot-cloudless-sky-blue above, with white wing margins, while the female is either dark blue or brown, with fiery-orange spots along the wing borders. The underside is very different from that of the Holly Blue, with large, bold black-and-white eye-spots and faded orange spots near the margins. This is a restless butterfly: a good way to go mad is to try to photograph one.

EGG (laid singly) > **LARVA** (caterpillar) > **PUPA** (underground, attractive to ants) > **ADULT**. The larval food plant is Bird's-foot Trefoil, and sometimes other peas and clovers. Overwinters as a caterpillar.

Females are immune to the charms of the bright blue colours of the males (they are more interested in scent). The colour is a warning signal to other males. The caterpillar produces a sugary substance from the honey gland on its seventh abdominal segment. Ants drum this segment when hungry.

TOP: *MALE UNDERSIDE*; BOTTOM: *MALE UPPERSIDE*

FACT FILE

ORDER Lepidoptera FAMILY Lycaenidae WINGSPAN 29–36mm SIMILAR SPECIES The Holly Blue has a different underside and tends to fly higher up, around bushes and trees. There are several other blues that are similar.

JAN | FEB | MAR | **APR** | **MAY** | **JUN** | **JUL** | **AUG** | SEP | OCT | NOV | DEC

HOLLY BLUE

Celastrina argiolus

This skittish butterfly is easy to see, but difficult to see well. It tends to fly high and eschews flowers, so the usual encounter is with a fast-flying, powder-blue butterfly circuiting around Holly bushes and other trees and shrubs, or ivy, and hardly ever settling. It prefers to drink honeydew well above the ground. However, due to its appearance early in the spring and its liking for gardens, parks and other suburban spaces it is easy to identify. It usually closes its wings at rest, and the underside is silvery-blue, sparsely seasoned with peppery spots. The upperside is brighter blue; the female shows black margins on the forewings. There are two generations, so it also flies in late summer.

WHERE TO FIND

Common throughout the region, except for most of Scotland and the far north of Scandinavia; patchy in Ireland.

EGG (on flower buds) > **LARVA** (caterpillar) > **PUPA** (in vegetation, often tended by ants) > **ADULT**. Overwinters as a pupa. Eggs laid in May give rise to larvae in June, pupae in July. Resulting adults lay eggs on ivy in August.

Most unusually, the larval food plants change with the season: Holly in spring, and ivy in summer and autumn.

UNDERSIDE

FACT FILE

ORDER Lepidoptera FAMILY Lycaenidae WINGSPAN 26–34mm SIMILAR SPECIES Common Blue (opposite), but there are also many other species of blue.

JAN	FEB	MAR	APR	**MAY**	**JUN**	**JUL**	AUG	SEP	OCT	NOV	DEC

LIME HAWK-MOTH

Mimas tiliae

WHERE TO FIND

Common and widespread, but only in the south of Scandinavia and the UK.

Hawk-moths are moth royalty. They are big, bold, colourful and fly with impressive, turbo-charged speed. There are a dozen or so species in Britain and more in Europe, and the Lime Hawk-moth is one of the most regular in suburban and urban gardens, where lime trees often grow. It is a handsome species, subtly pale green or brownish with bold, olive-green blotches across the forewing. The wings have crinkled edges and afford camouflage to the resting moth. It is sometimes seen on tree trunks and garden walls. Most hawk-moths have impressive larvae with a small horn at the rear end, looking like a tail.

EGG > **LARVA** (caterpillar, green with blue 'horn' at one end) > **PUPA** > **ADULT**. The larvae feed on limes and birches. It overwinters as a pupa below the ground. (Single generation, larvae June–September).

UPPERSIDE

★ A dozy Lime Hawk-moth once briefly interrupted a semi-final of the Wimbledon tennis championships.

FACT FILE

ORDER **Lepidoptera** FAMILY **Sphingidae** WINGSPAN **70–80mm** SIMILAR SPECIES **Other hawk-moths. The Poplar Hawk-moth *Laothoe populi* is another common hawk-moth, found further north.**

JAN	FEB	MAR	APR	**MAY**	**JUN**	**JUL**	**AUG**	SEP	OCT	NOV	DEC

ELEPHANT HAWK-MOTH

Deilephila elpenor

It is a remarkable fact that most people in our region have pink elephants in the garden and do not realize it. This wonderful and unmistakable moth, a mixture of jelly-green and pink as if it were an animate trifle dessert, or 'rhubarb and custard', is common, even in towns and cities. It is perhaps best seen using a special light trap to attract moths, although it is worth searching willowherbs in summer, and the moth also visits buddleia and honeysuckles. The English name derives from its equally outlandish caterpillar, which has four extraordinary 'eye'-spots and a protruding front that could fancifully be compared to an elephant's trunk.

WHERE TO FIND

Locally common throughout the region, up to the southern half of Scandinavia and Finland.

EGG (singly or pairs, on leaves) > **LARVA** (caterpillar) > **PUPA** > **ADULT**. The larvae feed on willowherbs. Overwinters as a pupa.

The larvae can stand immersion in water. The moth has exceptional colour vision in low light.

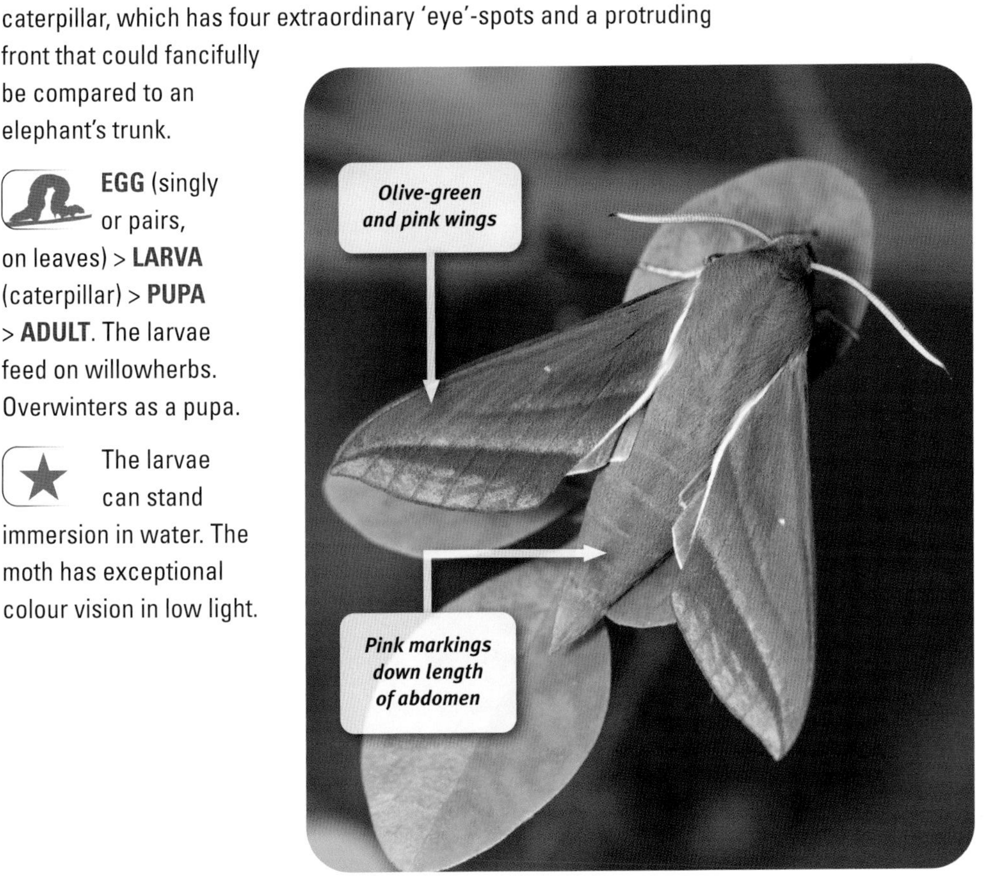

UPPERSIDE

FACT FILE

ORDER Lepidoptera FAMILY Sphingidae WINGSPAN 62–72mm SIMILAR SPECIES There is a smaller and quite similar species, but nothing else remotely resembles it.

JAN FEB MAR **APR MAY JUN JUL AUG SEP OCT NOV DEC**

HUMMINGBIRD HAWK-MOTH

Macroglossum stellatarum

WHERE TO FIND

Rarely common but can turn up literally anywhere in the region and in any month, even in the Arctic Circle.

Hummingbirds are only found in the Americas, but every year people spot this hawk-moth in their garden and are convinced that one of the jewel-like birds has arrived here. This insect is our nearest equivalent. Its tendency to hover with audibly whirring wings at trumpet-like flowerheads, using its long proboscis to drink nectar, is an impressive example of convergent evolution. With its grey body, orange hindwings and black-and-white rear end, it is unmistakable.

EGG > **LARVA** (caterpillar, green with purplish-red horn) > **PUPA** (cocoon in foliage of food plant or litter below) > **ADULT**. The larvae feed on bedstraws. It overwinters as an adult, hibernating in tree holes and outbuildings.

It sometimes flies in the rain, unusually for a moth.

TOP & ABOVE: *UPPERSIDE*

FACT FILE

ORDER **Lepidoptera** FAMILY **Sphingidae** WINGSPAN **50–58mm** SIMILAR SPECIES **Unique.**

JAN FEB MAR **APR MAY JUN JUL AUG SEP OCT** NOV DEC

GARDEN CARPET

Xanthorhoe fluctuata

This is one of many butterfly-like moths that hold their wings open at rest. In this species the main black cross-band stops half-way as if the central part has been rubbed out.

EGG > **LARVA** (caterpillar) > **PUPA** (underground) > **ADULT**. The larvae feed on members of the cabbage family, such as Shepherd's Purse. It overwinters as a pupa.

ABOVE: UPPERSIDE

WHERE TO FIND
Common throughout the region.

FACT FILE

ORDER **Lepidoptera** FAMILY **Geometridae** WINGSPAN **27–31mm** SIMILAR SPECIES **There are many similar species, such as the Common Carpet *Epirrhoe alternata*, in which the dark bar goes right across the wing.**

JAN FEB MAR APR MAY **JUN JUL AUG** SEP OCT NOV DEC

YELLOW SHELL

Camptogramma bilineata

Providing a classic example of the intricacy of moth patterns, the Yellow Shell looks like a dull yellow nonentity when disturbed by day. But its complex display of lines, waves and panels helps to camouflage it and disrupt its overall shape.

EGG > **LARVA** (caterpillar) > **PUPA** (in loose earth) > **ADULT**. The larvae feed on various food plants, including bedstraws and docks. The larva emerges from July, then overwinters.

ABOVE: UPPERSIDE

WHERE TO FIND
Very common throughout the region.

FACT FILE

ORDER **Lepidoptera** FAMILY **Geometridae** WINGSPAN **28–32mm** SIMILAR SPECIES **Many other moths are of similar size and shape, and a few are yellowish.**

JAN FEB MAR APR **MAY** **JUN** **JUL** **AUG** **SEP** **OCT** **NOV** DEC

BLOOD-VEIN

Timandra comae

WHERE TO FIND

Widespread and common up to the southern half of Scandinavia. Scarce in Ireland and Scotland.

Some moths are as easy to identify as butterflies, and the gorgeous Blood-vein is an example. No other species has the dark 'vein' running straight across the middle of the body, in combination with the shocking pink trailing edge. The cross-line disrupts the shape of the insect at rest so that predators will not recognize it as a moth. This is a very common species that occurs in all kinds of gardens and can be disturbed from vegetation during the day.

EGG > **LARVA** (caterpillar) > **PUPA** (among plant debris on the ground) > **ADULT**. The larvae feed on various food plants, especially docks and sorrel. It overwinters as a larva. There are usually two generations of adult moths, in May–July and July–September.

The larvae are badly affected by high nitrogen levels in the host plant, making the species vulnerable to fertilizers in garden soil.

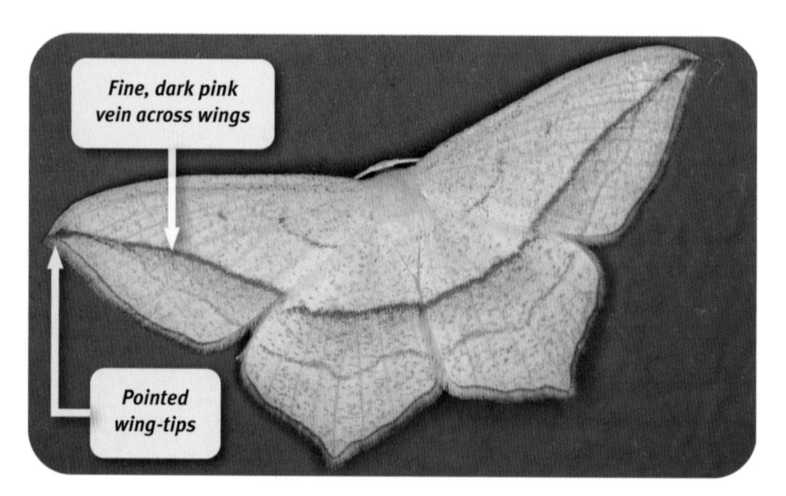

UPPERSIDE

FACT FILE

ORDER Lepidoptera FAMILY Geometridae WINGSPAN 30–34mm SIMILAR SPECIES There are dozens of moths with a similar whitish ground colour, often with cross-lines, but none like this one. There is also the Small Blood-vein Moth *Scopula imitaria*, but in this species the 'vein' is more of a smudged charcoal line as opposed to a bloody vein.

JAN FEB MAR APR MAY **JUN JUL AUG SEP OCT** NOV DEC

WILLOW BEAUTY

Peribatodes rhomboidaria

This is the sparrow among moths: brown, common, overlooked and underappreciated. To be honest, the English name could only have been bestowed by a besotted lover, because any 'beauty' is definitely in the eye of the beholder. It perches elegantly enough with its wings open, butterfly style, but the pale brown ground colour, with numerous darker brown cross-lines and blotches, is made to be overlooked, not admired. Often appearing at windows and resting on walls, this nocturnal moth is one of the most common found at moth traps. The caterpillars feed on a wide variety of plants.

EGG > **LARVA** (caterpillar) > **PUPA** (cocoon among plant debris) > **ADULT**. The larvae feed on various food plants, including privet, honeysuckles and ivy. It overwinters as a larva. In the south there may be two generations, the second beginning in August.

This is a notorious pest of vineyards in southern Europe.

WHERE TO FIND

Very common throughout the region.

UPPERSIDE

FACT FILE

ORDER Lepidoptera FAMILY Geometridae WINGSPAN 40–48mm SIMILAR SPECIES There are many similar moths in the garden. They include the Mottled Beauty *Alcis repandata*, which can be differentiated by the dark wavy cross-line across the wings – the Willow Beauty has two very faint cross-lines.

JAN FEB MAR **APR MAY JUN JUL AUG SEP OCT** NOV DEC

BRIMSTONE MOTH

Opisthograptis luteolata

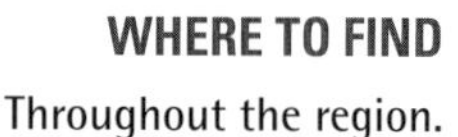

WHERE TO FIND

Throughout the region.

Yes, there is a Brimstone Moth as well as a Brimstone butterfly. This is a very common garden moth that is often attracted to lighted windows and can readily be seen on a torchlight foray by night. At rest it holds its wings flat and open like a butterfly. The buttery-yellow colour, brown blotches and square marks are all unique, although there are many whitish and cream-coloured moths of similar size.

EGG > **LARVA** (caterpillar) > **PUPA** > **ADULT**. The larvae feed on various food plants, including hawthorn, blackthorn and plums. It sometimes overwinters as a larva and sometimes as a pupa. There may be two or three generations.

The Brimstone Moth caterpillar strongly resembles a twig and positions itself carefully to fool any passing predator. This is known as masquerading.

UPPERSIDE

FACT FILE

ORDER Lepidoptera FAMILY Geometridae WINGSPAN 33–46mm SIMILAR SPECIES It looks like a butterfly, but is nocturnal.

JAN FEB MAR APR **MAY JUN JUL AUG SEP** OCT NOV DEC

LIGHT EMERALD

Campaea margaritaria

There are very few green moths or butterflies, but this is one you might see – and a beautiful one, too, with its delicate shy green colour, a darker green mid-band edged with white, and a curious reddish mark on the tip of the forewing, which breaks up the moth's shape to fool a potential predator. The green soon fades to whitish. The moth flies from dusk onwards and is attracted to light, so it is a species that can become familiar to gardeners, who might also flush it from the undersides of leaves. The caterpillar is unusual in having hairs all along its underside.

EGG (laid in clusters, with reddish dots, on undersides of leaves) > **LARVA** (caterpillar, feeds on leaves) > **PUPA** (on underside of leaf) > **ADULT**. It overwinters as a larva lying along a stem. There are two generations. The overwintering larvae mature in May–July, and lay eggs that give rise to the new generation in August to the beginning of October.

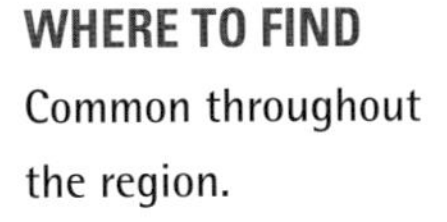

WHERE TO FIND

Common throughout the region.

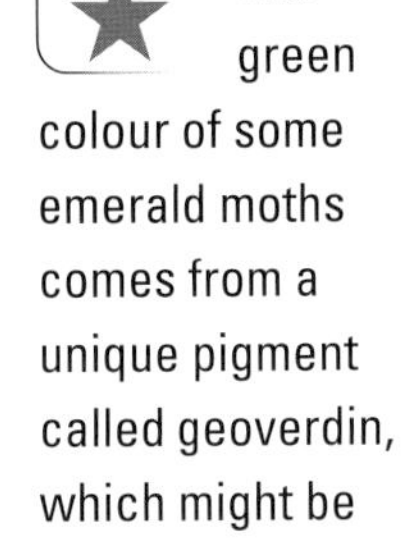

The green colour of some emerald moths comes from a unique pigment called geoverdin, which might be derived from the chlorophyll of leaves.

UPPERSIDE

FACT FILE

ORDER Lepidoptera FAMILY Geometridae WINGSPAN 42–54mm SIMILAR SPECIES There are several species of emerald moth, but the others are mostly darker green and lack the red mark at the wing-tip in this species.

JAN	FEB	MAR	APR	**MAY**	**JUN**	**JUL**	**AUG**	SEP	OCT	NOV	DEC

BUFF-TIP

Phalera bucephala

WHERE TO FIND
Common throughout the region.

There is probably no other moth that has made more children fall in love with the mini-beasts in their garden than the Buff-tip. A benign, chunky moth, it is scientifically known as a 'twig mimic', which means that its nature is to sit still by day and let its remarkable camouflage protect it. The resemblance to a birch twig, especially the moth's pale 'face', which resembles a broken twig end, is incredible. If you come across it, perhaps at an organized 'moth-trapping' event (p. 7), the moth will contentedly sit on your finger. The moth flies at night throughout the warmest months of the year.

EGG (laid in batches on undersides of leaves) > **LARVA** (caterpillar, black with yellow stripes, living in small groups) > **PUPA** (in cell in soil) > **ADULT**. The larvae feed on leaves of willows, birches, oaks and many other trees. It overwinters as a pupa underground.

Mating can go on for hours. A male may so occupy a female that by the time he is finished she is no longer chemically attractive to other males.

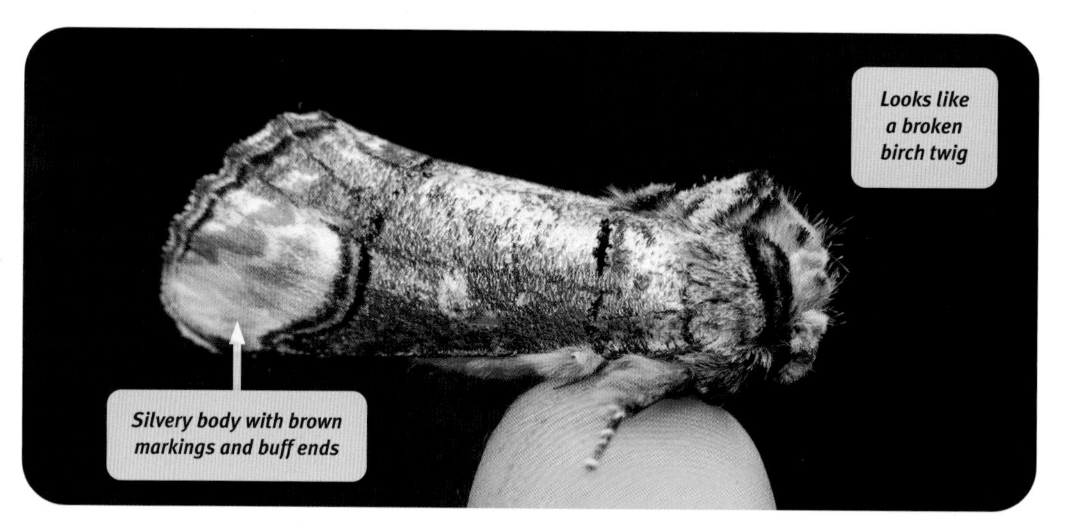

UPPERSIDE

FACT FILE

ORDER Lepidoptera FAMILY Notodontidae WINGSPAN 55–68mm SIMILAR SPECIES This is a unique moth that will not be confused with anything else.

JAN FEB MAR **APR MAY JUN JUL AUG SEP** OCT NOV DEC

RUBY TIGER

Phragmatobia fuliginosa

WHERE TO FIND
Common throughout the region.

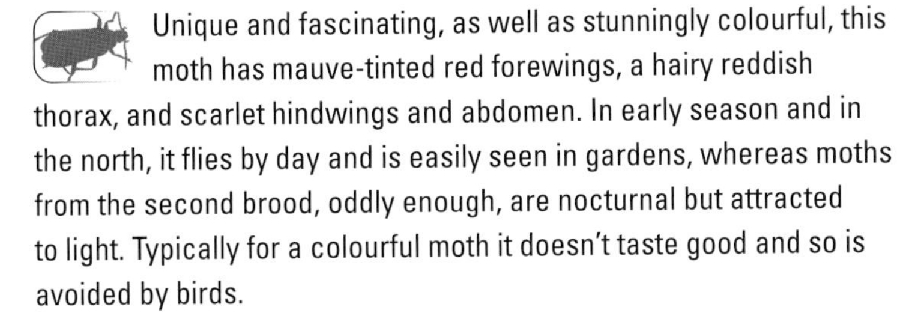

Unique and fascinating, as well as stunningly colourful, this moth has mauve-tinted red forewings, a hairy reddish thorax, and scarlet hindwings and abdomen. In early season and in the north, it flies by day and is easily seen in gardens, whereas moths from the second brood, oddly enough, are nocturnal but attracted to light. Typically for a colourful moth it doesn't taste good and so is avoided by birds.

EGG (laid on leaf surfaces) > **LARVA** (caterpillar feeds on a wide range of plants, including plantains and docks) > **PUPA** (cocoon on or below ground) > **ADULT**. It overwinters as a larva. (First brood mid-April–June, second mid-July–early September.)

Makes clicking sounds when it detects the ultrasonic calls of bats. The clicks send a message that the moth is distasteful.

LEFT: *LATERAL VIEW*; RIGHT: *UPPERSIDE*

FACT FILE

ORDER Lepidoptera FAMILY Erebidae WINGSPAN 28–38mm SIMILAR SPECIES None.

JAN FEB MAR APR MAY **JUN** **JUL** **AUG** **SEP** OCT NOV DEC

GARDEN TIGER

Arctia caja

WHERE TO FIND

Common throughout the region, right up into northern Scandinavia, but nowhere near as abundant as it once was and is declining alarmingly.

If ever a species exploded the myth that moths were dull and boring, it would have to be this one. It is gorgeous and unmistakable, although individuals vary in colour scheme. It was once a moth that anyone might come across, perhaps when disturbing it from vegetation by day, or seeing it resting on a house wall or coming to a lighted window at night, but it is much less common now and is easiest to see when attending a summer moth-trapping session. The caterpillar is also famous, brown and hairy, and known colloquially as a 'woolly bear'. It has an odd habit of wandering around on open ground, such as paths, by day. It is a garden gem to make anyone fall in love with nature.

EGG (on leaf surfaces) > **LARVA** (caterpillar) > **PUPA** (cocoon on or below the ground; pupates in spring) > **ADULT**. Adults feed on nectar. Larvae feed on nettles, docks and many other plants. It overwinters as a larva, in leaf litter or plant debris.

The larva ingests a battery of toxic compounds from the plants it eats. Not only can it break these down into non-toxic form, but it can also use them in slightly altered form for defence against predators.

LEFT: *ADULT*; RIGHT: *LARVA*

FACT FILE

ORDER Lepidoptera FAMILY Erebidae WINGSPAN 50–78mm SIMILAR SPECIES A few related species approach the same pattern, but are smaller and day flying.

JAN | FEB | MAR | APR | **MAY** | **JUN** | **JUL** | **AUG** | SEP | OCT | NOV | DEC

CINNABAR

Tyria jacobaeae

It is unusual for a caterpillar to outshine a moth, particularly a very beautiful one, but such is the familiarity of the yellow-and-black-banded Cinnabar moth larva, chewing its way communally through clumps of ragwort, that its profile is higher than its stunning but easily overlooked adult stage. The caterpillars come out en masse in June and are still obvious in September. The adult moth, which flies by day, is discreet and weak flying. At rest the wings are half closed, but the scarlet underwings, together with streaks and spots in the upperwings, are still obvious.

WHERE TO FIND
Common, but absent from northern Scotland and all but the extreme south of Norway and Sweden.

EGG (batches of 30–60) > **LARVA** (caterpillar, yellow and black) > **PUPA** (below the ground) > **ADULT**. The larvae feed communally on ragwort, and also on groundsel. It overwinters as a pupa.

The larvae absorb poisons from the food plant and are dangerously unpalatable to predators.

LEFT: *ADULT*; RIGHT: *LARVA*

FACT FILE

ORDER Lepidoptera FAMILY Erebidae WINGSPAN 35–45mm SIMILAR SPECIES Look out for some brilliantly scarlet day-flying moths called burnets, although these are unusual in gardens. Burnet moths have only red spots on their wings, whereas Cinnabars have red spots and red bars.

JAN | FEB | MAR | APR | **MAY** | **JUN** | **JUL** | **AUG** | **SEP** | **OCT** | **NOV** | DEC

LARGE YELLOW UNDERWING

Noctua pronuba

WHERE TO FIND
Abundant everywhere throughout the region.

This is an elongated, chunky moth that looks dull and boring until it flies, which it often does when disturbed from grass or other vegetation in late summer, then showing its yellow wing-flashes. It is one of our most common species, and almost any solidly built, fast-moving moth that you see in August is likely to be this ubiquitous species. The yellow underwings help to distract predators that come across it. It is one of those moths that might fly in through a bedroom window, attracted to light. It also feeds at night from any number of garden plants. The population is swollen by southern immigrants in some summers.

EGG > **LARVA** (caterpillar, lives underground and feeds at night) > **PUPA** (below the ground in spring) > **ADULT**. The larvae feed on many different plants, including docks. It overwinters as a larva.

Incredibly, it seems to be able to navigate using the rotation of the stars when migrating at night.

LEFT & RIGHT: *UPPERSIDE*

FACT FILE

ORDER Lepidoptera FAMILY Noctuidae WINGSPAN 50–60mm SIMILAR SPECIES There are some other closely related species with yellow underwings, but this is the largest.

JAN | FEB | MAR | APR | **MAY** | **JUN** | **JUL** | **AUG** | SEP | OCT | NOV | DEC

HEART AND DART

Agrotis exclamationis

The midsummer night's moth, this is often the most abundant species on the shortest nights of the year. It visits flowers at night and comes to lighted windows, and abounds at moth traps. It is refreshingly easy to identify, despite being one of a multitude of medium-sized, chunky moths. The wing pattern is certainly distinctive, with heart shapes and streaks (darts) in the mid-wing, but to a moth enthusiast the diagnostic feature is the unique black collar behind the moth's snout.

EGG > **LARVA** (caterpillar) > **PUPA** (below ground; pupates in spring) > **ADULT**. The larvae feed on many different plants. It overwinters as a larva.

This moth often migrates. It uses the moon as its primary cue, but the moon moves across the sky. Incredibly, these moths, with their tiny brains, use their geomagnetic sense to compensate for the moon's movements.

WHERE TO FIND

Abundant throughout the region.

UPPERSIDE

FACT FILE

ORDER Lepidoptera FAMILY Noctuidae WINGSPAN 35–44mm SIMILAR SPECIES There are dozens of species of similarly coloured and shaped moths.

JAN | FEB | MAR | APR | MAY | JUN | JUL | AUG | SEP | OCT | NOV | DEC

ANGLE SHADES

Phlogophora meticulosa

WHERE TO FIND

Common in every corner of the region.

This is a surprisingly distinctive moth with an unusual combination of olive-green, pale pink and whitish blotches and triangles. Not only does the overall pattern match a crumpled-up leaf; so do the crinkled edges to the wings, which are unique. The Angle Shades is a common species that can travel a long way to reach a garden, since in the summer many individuals migrate up from southern Europe and swell the numbers already present. It is quite often seen by day on walls, fences and vegetation, and at night it is attracted to light. It visits flowers at night.

EGG (in batches on leaves and grass) > **LARVA** (caterpillar) > **PUPA** (cocoon just below the soil's surface) > **ADULT**. The larvae feed on a wide variety of plants, including nettles and docks. It overwinters as a larva.

It is strongly migratory and there is even a population in the Azores, 1,400km out in the Atlantic Ocean.

LEFT: *ADULT*; RIGHT: *LARVA*

FACT FILE

ORDER Lepidoptera FAMILY Noctuidae WINGSPAN 45–52mm SIMILAR SPECIES The shape is unique. If this species were bigger, it would not look out of place as a hawk-moth.

JAN | FEB | **MAR** | **APR** | **MAY** | JUN | JUL | AUG | SEP | OCT | NOV | DEC

HEBREW CHARACTER

Orthosia gothica

Spring may still be far away when the first Hebrew Characters emerge from their underground cocoons and begin to fly, feeding on sallow and other early blossoms during the long nights. A few will even fly in autumn and winter. This is our only early-flying moth with such a distinctive black mark on its forewing, so it is easily identified. Go out on a warm evening in March and you might well spot one in the torchlight or flying towards the lights at a window. It is common absolutely everywhere.

WHERE TO FIND
Common throughout the region and found in any habitat.

EGG > **LARVA** (caterpillar, feeds on leaf buds, then leaves) > **PUPA** (below the ground) > **ADULT** (remains in cocoon for some time). The larvae feed on various plants, including birch and oak leaves. It overwinters as a pupa.

This moth's blood contains antifreeze, allowing it to survive in very cold weather.

UPPERSIDE

FACT FILE

ORDER Lepidoptera FAMILY Noctuidae WINGSPAN 30–40mm SIMILAR SPECIES There are a number of similar moths. One, the Setaceous Hebrew Character *Xestia c-nigrum*, starts in May and continues into September.

JAN FEB MAR APR MAY **JUN JUL AUG SEP** OCT NOV DEC

BURNISHED BRASS

Diachrysia chrysitis

WHERE TO FIND

Common throughout the region.

If there were no nettles, there would be no gorgeous Burnished Brass, a metallic beauty of a moth that takes your breath away. The light reflects on that brassy loveliness, sometimes producing pure gold, at times with a greenish tinge. It seems impossible that a moth could be so startling, yet fly at night in the darkness and somehow hide by day. It sometimes emerges at dusk and begins feeding on the nectar of flowers. The best way to see one is to set up a moth trap on a hot summer night in the garden.

EGG (green, vaguely resembling a jellyfish) > **LARVA** (caterpillar; feeds at night and hides at base of nettle by day) > **PUPA** (inside loose cocoon underneath a leaf) > **ADULT**. The larvae usually feed on nettles. It overwinters as a larva on the ground. There may be two generations a year, with eggs laid by adults in June–July producing adults in August–October of the same year.

This common moth may actually be two species, with slight varation in the markings. They might also contain different sex-attractant chemicals.

UPPERSIDE

FACT FILE

ORDER Lepidoptera FAMILY Noctuidae WINGSPAN 34–44mm SIMILAR SPECIES There are several similar species, mostly found on the Continent.

JAN | FEB | MAR | APR | **MAY** | **JUN** | **JUL** | AUG | SEP | OCT | NOV | DEC

YELLOW-BARRED LONGHORN

Nemophora degeerella

There are just a few species of moth with greatly extended antennae, and this is by far the smartest and most colourful, with the golden-yellow band across its wing. It is mostly found in damp woodland and hedgerows, but will visit wilder gardens. The length of the male's antennae is simply extraordinary, up to five times as long as the body, the longest of any British moth or butterfly, while in the female the antennae are more modest (as long as the wing) and thickened at the base. This is a day-flying moth.

EGG > **LARVA** (on leaf litter) > **PUPA** (in leaf litter) > **ADULT**. It overwinters as a larva in a cocoon in leaf litter.

Males are often seen 'dancing' in groups over sunlit bushes and tree branches, well into the evening.

WHERE TO FIND

Common and widespread, but rare in Ireland and patchy in Scotland.

UPPERSIDE

FACT FILE

ORDER Lepidoptera FAMILY Adelidae BODY LENGTH 17–23mm SIMILAR SPECIES There are several other longhorn moths, but none have the gold wing-band.

JAN	FEB	MAR	APR	**MAY**	**JUN**	**JUL**	AUG	SEP	OCT	NOV	DEC

GREEN OAK TORTRIX

Tortrix viridana

WHERE TO FIND

Common and widespread, but only in the extreme south of Scandinavia.

Very few moths are green and, amazingly, this is the only 'micro-moth' (that is, a tiny moth that you could blow away) among the 1,500 British species to have plain green wings. As such it is unmistakable. However, you are just as likely to encounter it under an oak tree in midsummer, when the caterpillar descends by a silken thread and may hang in the warm air. This is often an abundant moth, and in some years the larvae can defoliate entire trees.

UPPERSIDE

EGG (in leaf buds of oak and other trees) > **LARVA** (feeds on new leaves) > **PUPA** (in folded or rolled-up oak leaf) > **ADULT**.

The larvae hatch before the leaves have burst, to take advantage of the time when the leaves are most nutritious. They are sometimes too early and rest on bare branches, but 50 per cent of the population can survive for 12 days without feeding.

FACT FILE

ORDER **Lepidoptera** FAMILY **Tortricidae** WINGSPAN **17–24mm** SIMILAR SPECIES **There are a few green moths, but the size and shape of this one is distinctive.**

JAN	FEB	MAR	APR	**MAY**	**JUN**	**JUL**	**AUG**	**SEP**	OCT	NOV	DEC

SMALL MAGPIE

Anania hortulata

This is a fragile-looking moth that holds its wings out flat at rest, although the forewing edge is always down at an angle. As its name implies, it is boldly black and white, but the head and thorax are stained orange-yellow.

WHERE TO FIND
Occurs commonly throughout the region.

EGG > **LARVA** (caterpillar) > **PUPA** (in cocoon) > **ADULT**. Larva feeds on a nettle leaf that it rolls to make a hiding place; also some other herbs such as mint. It overwinters as a larva.

ABOVE: UPPERSIDE

FACT FILE

ORDER **Lepidoptera** FAMILY **Crambidae** WINGSPAN **33–35mm** SIMILAR SPECIES **A handful of other black-and-white moths occur in gardens, but most are larger than this species.**

JAN	FEB	**MAR**	**APR**	**MAY**	**JUN**	**JUL**	**AUG**	**SEP**	OCT	NOV	DEC

SMALL PURPLE AND GOLD

Pyrausta aurata

This gorgeous moth would be a great favourite were it not for its diminutive size. It flies during the day like a butterfly, but it never sleeps and also visits flowers at night. It is easily overlooked among the summer insect community.

WHERE TO FIND
Common and widespread, but not found in Ireland and scarce in Scotland.

EGG > **LARVA** (caterpillar, feeds on leaf first, then flowers, where it spins a web) > **PUPA** > **ADULT**. Larva feeds particularly on mints and marjoram.

ABOVE: UPPERSIDE

FACT FILE

ORDER **Lepidoptera** FAMILY **Pyralidae** WINGSPAN **15–18mm** SIMILAR SPECIES **There are several closely related species, but this one is the most colourful.**

JAN | FEB | MAR | APR | MAY | **JUN** | **JUL** | **AUG** | **SEP** | **OCT** | NOV | DEC

MOTHER-OF-PEARL

Patania ruralis

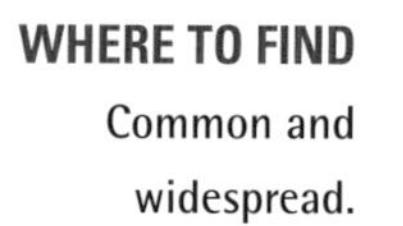

WHERE TO FIND
Common and widespread.

At first sight this might look like 'just another moth', but in the right light it reveals itself to have a lustrous pearly sheen that gives it its name, showing that it is always worth taking a closer look at all our insects. In summer you can almost always disturb a Mother-of-pearl from a nettle patch or other vegetation. It also comes out early, at dusk, so again is one of our easiest moths to see.

UPPERSIDE

EGG > **LARVA** (caterpillar) > **PUPA** (in cocoon) > **ADULT**. Larva feeds on a nettle leaf that it rolls to make a hiding place. It overwinters as a larva in a silk cocoon.

It has an overseas career as a serious pest of soya bean crops.

FACT FILE

ORDER **Lepidoptera** FAMILY **Crambidae** WINGSPAN **33–37mm** SIMILAR SPECIES **The pearly sheen is unique.**

JAN | FEB | MAR | APR | **MAY** | **JUN** | **JUL** | AUG | SEP | OCT | NOV | DEC

BLACK-BACKED SAWFLY

Tenthredo mesomela

A green wasp-like insect common in rich herbage. It spends much time hunting for pollen and nectar from flowers, with a sideline in predation on small insects. Bees and wasps are descended from sawflies.

WHERE TO FIND

Widespread and common.

EGG (laid inside plant stem) > **LARVA** (caterpillar-like, feeds on leaves, mainly of buttercups) > **PUPA** > **ADULT**.

FACT FILE

ORDER Hymenoptera FAMILY Tenthredinidae BODY LENGTH 10–13mm SIMILAR SPECIES There are some similar sawflies, especially the psychedelically green Green Sawfly *Rhogogaster viridis*.

JAN | FEB | MAR | APR | **MAY** | **JUN** | **JUL** | **AUG** | **SEP** | **OCT** | NOV | DEC

TURNIP SAWFLY

Athalia rosae

The sawflies are one of the great unknowns of the garden; there are dozens of species. The adults can be seen on flowers, especially umbellifers and dandelions. They fly slowly and are easily overlooked.

WHERE TO FIND

Recorded over much of the region, but absent from Ireland and northern Scotland.

EGG (laid inside plant stem, female has saw-like appendage for cutting into it) > **LARVA** (black, caterpillar-like, feeds on host plant) > **PUPA** > **ADULT** (brief stage, just a few days.)

FACT FILE

ORDER Hymenoptera FAMILY Tenthredinidae BODY LENGTH 7–8mm SIMILAR SPECIES There are many. Sawflies resemble the great multitude of wasp-like creatures in the garden.

JAN FEB MAR APR **MAY JUN JUL AUG SEP** OCT NOV DEC

JAVELIN WASP

Gasteruption jaculator

WHERE TO FIND

Patchily distributed but probably overlooked. In the UK, scarce in Wales and Scotland. Apparently absent from Ireland, the Netherlands, Denmark and Norway.

Arguably the slimmest of all garden insects, this remarkable wasp is a parasite – or strictly speaking a parasitoid – on solitary bees and wasps. Parasites are annoying, but parasitoids are fatal. In this case the female lays eggs inside a nesting chamber carrying bee or wasp grubs, then her eggs hatch first and her own larvae proceed to eat their larval host, as well as detritus from the bottom of the nest. The female of this species is exceptional for the length of the 'tail' – actually the ovipositor – which is inserted into the burrow. There are a surprising number of parasitic wasp species, many poorly known, even the garden ones.

EGG (laid in nest of solitary bee or wasp) > **LARVA** (grub, in host burrow) > **PUPA** (inside host burrow) > **ADULT** (feeds on nectar and pollen). A parasitoid. Larva feeds on grubs of host.

This wasp uses 'vibrational sounding' to detect a host's nest – similar to knocking on a wall to find a secret passage. It is able to manipulate the end of the tip of its ovipositor.

FACT FILE

ORDER **Hymenoptera** FAMILY **Gasteruptiidae** BODY LENGTH **10–18mm** SIMILAR SPECIES **Probably the slimmest of all the parasitic wasps, but check ichneumons. It looks very similar to ichneumons but is different because its long ovipositor is generally used for probing and laying in open areas, as opposed to drilling into closed spaces, as ichneumons do.**

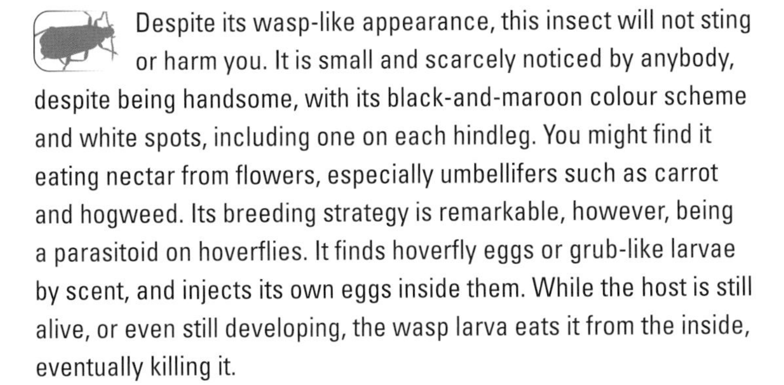

SPOT-LEGGED ICHNEUMON

Diplazon laetatorius

Despite its wasp-like appearance, this insect will not sting or harm you. It is small and scarcely noticed by anybody, despite being handsome, with its black-and-maroon colour scheme and white spots, including one on each hindleg. You might find it eating nectar from flowers, especially umbellifers such as carrot and hogweed. Its breeding strategy is remarkable, however, being a parasitoid on hoverflies. It finds hoverfly eggs or grub-like larvae by scent, and injects its own eggs inside them. While the host is still alive, or even still developing, the wasp larva eats it from the inside, eventually killing it.

WHERE TO FIND

Probably widespread, but consistently overlooked; found worldwide, even in Iceland.

EGG (inside egg or larva of hoverfly) > **LARVA** (develops inside living host) > **PUPA** > **ADULT**.

There are more than 2,500 species of ichneumon in Britain, but many people have never heard of them. Female parasitic wasps can control the sex of their offspring by choosing whether the eggs are fertilized, leading to females, or not, which leads to males.

FACT FILE

ORDER Hymenoptera FAMILY Ichneumonidae BODY LENGTH 4–7mm SIMILAR SPECIES There are many similar insects in gardens and elsewhere. Ichneumons are the great unknown of British entomology.

JAN FEB MAR **APR MAY JUN JUL AUG SEP OCT** NOV DEC

RUBY-TAILED WASPS

Chrysididae spp.

WHERE TO FIND
Widespread and fairly common throughout the region, apart from the most northerly areas.

This is one of the loveliest of all insect species in the garden – the dazzling, glittering body is an exquisite metallic bottle-green and intense ruby in colour, and seeing one takes your breath away. Furthermore, as far as humans are concerned, it is harmless and cannot sting. Look for it on a sunny day investigating holes on surfaces; it also eats nectar from flowers. However, it has a dark side. It is a parasitoid (a fatal parasite) of solitary wasps and bees, laying its eggs in their nests. The female spends much time investigating used burrows, in which it lays an egg.

EGG (in nest of host) > **LARVA** (develops in nest of host; eats food provided for host young, then will eat the grub itself) > **PUPA** (in nest) > **ADULT**.

The body is specially toughened, and when threatened by a host the wasp can roll itself into a defensive ball. The jaws and stings of wasps and bees are unable to penetrate its body armour.

FACT FILE

ORDER Hymenoptera FAMILY Chrysididae BODY LENGTH 9mm SIMILAR SPECIES There are several similar species.

JAN | FEB | MAR | APR | MAY | JUN | JUL | AUG | SEP | OCT | NOV | DEC

RED ANT

Myrmica rubra

The Red Ants of the garden need to be treated carefully. If provoked or deliberately handled, they are able to inflict a painful bite. They are larger than Yellow Meadow or Small Black Ants (pp. 142 and 143), and they live in nests of about 1,000 workers. They crawl all over the lawn, beds, paths and surrounding low vegetation, and the colonies are often under stones and flowerpots, as well as in the soil. Their food includes aphids, small worms, flies, and also honeydew (secretions from aphids) and nectar. Their nests are often close together in 'super-colonies'.

EGG (tended by workers in nests) > **LARVA** (as egg) > **PUPA** (as egg) > **ADULT**. There are usually many queens in a colony (15 queens per 1,000 workers), which produce workers (all female, wingless), males and future queens. In late summer the males and newly hatched queens embark on mating flights.

When a fungal infection breaks out in a nest, workers somehow know to decrease their interactions with other colony members.

WHERE TO FIND

Generally common throughout the region, except in northern Scandinavia, Scotland and Ireland (where it is replaced by similar species). Often found in places with high rainfall, but below 1,500m.

ABOVE: *WORKER*

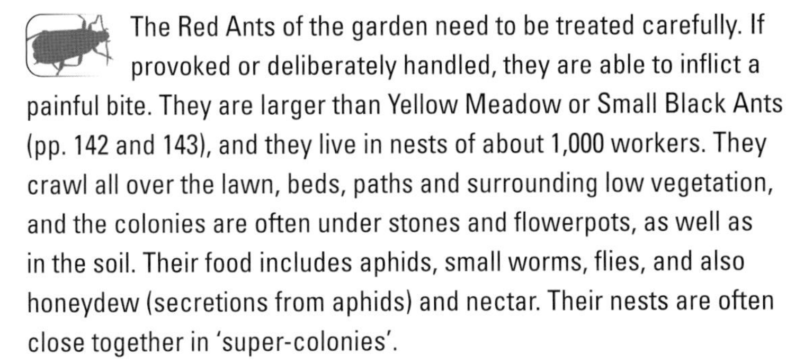

FACT FILE

ORDER Hymenoptera FAMILY Formicidae BODY LENGTH 4–8mm SIMILAR SPECIES There are several closely similar species of red ant. Wood Ants are larger.

JAN | FEB | MAR | APR | MAY | JUN | JUL | AUG | SEP | OCT | NOV | DEC

YELLOW MEADOW ANT

Lasius flavus

WHERE TO FIND
Common throughout the region, except in far northern Scandinavia.

It is an ant and it is yellow – and that completes the identification lesson for this very common garden ant. It lives in large colonies and these are the ones that make those circular mounds, often covered in grass, that you often find and trip over in meadows. The mounds may have been built up over many years by countless generations of ants, and occupied ones may host many thousands of individuals. On lawns, Yellow Meadow Ants live below ground and have a most exotic diet – they eat the honeydew secreted by aphids that live on roots.

EGG (tended by workers in nest) > **LARVA** (as egg) > **PUPA** (as egg) > **ADULT**. There may be several queens in a big colony and only they reproduce, hatching workers (all female, wingless), males and future queens. In late summer the males and newly hatched queens embark on mating flights.

A mound made by these ants can be as tall as 60cm.

FACT FILE

ORDER Hymenoptera FAMILY Formicidae BODY LENGTH 3–5mm SIMILAR SPECIES Smaller than the other common garden ants. The males and queens are browner than the workers, especially on the head.

JAN | FEB | MAR | APR | MAY | JUN | JUL | AUG | SEP | OCT | NOV | DEC

SMALL BLACK ANT

Lasius niger

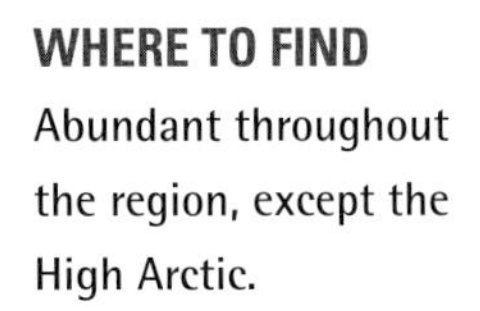

It is impossible to miss this indomitable garden insect – after all, if you do not detect it running over paving slabs, soil and grass, it will probably find you, working its way uninvited into the kitchen in search of sweet food. It is really dark brown rather than black. The colonies are found in the soil under rocks and tree stumps, and in common with other ants there are divisions of labour among castes. There is just one queen, the only reproductive member of the colony. Workers find food for the grubs and carry out colony maintenance and defence. They will eat many types of food, including honeydew from aphids.

WHERE TO FIND
Abundant throughout the region, except the High Arctic.

EGG (white, tended by workers in nest) > **LARVA** (grub, three instars, as egg) > **PUPA** (in cocoon, as egg) > **ADULT**. There is a single queen in a colony, and it produces the new workers (all female, wingless), males and new queens. In late summer the males and newly hatched queens embark on mating flights.

These animals are remarkably long lived. Workers may live for four years, queens for 15 years. Every year there is a 'flying ant day' when these ants perform their mating flights all at the same time, often flying into people's faces and seemingly filling the air. When a worker finds food, it leaves a chemical trail that others can follow.

FACT FILE

ORDER Hymenoptera FAMILY Formicidae BODY LENGTH 4–5mm SIMILAR SPECIES This is by far the most common ant of its colour that is found in gardens.

JAN FEB **MAR APR MAY JUN JUL AUG SEP OCT NOV** DEC

COMMON WASP

Vespula vulgaris

WHERE TO FIND
Common everywhere in the region, except for Arctic areas.

Always an insect to induce mass hysteria, the sweet-toothed Common Wasp likes to share human summer picnics. It rarely stings if left alone, but can be provoked by rapid movements. In the spring it is barely noticed, the workers spending most of their time feeding on the smaller invertebrates of the garden and acting as natural pest controllers. Wasps are also vital pollinators. They live in paper-bark nests (made from wood and saliva), first constructed by the queen early in the spring, underground or sometimes in sheds. The queen produces hundreds of eggs a day in spring, having mated before hibernation. Thousands of workers are then produced during summer, which feed the grubs and look after the nest. Later on, males and virgin queens emerge. When the queen dies the workers run amok, and it is then that they 'annoy' humans.

EGGS (in nest of social colony) > **LARVA** (grub) > **PUPA** (in nest in colony) > **ADULT**. The larvae are fed by the workers on insect flesh. It overwinters as an adult queen.

Wasps are early risers. They must forage as soon as it gets light to feed the larvae, which might otherwise starve. They also forage for dew.

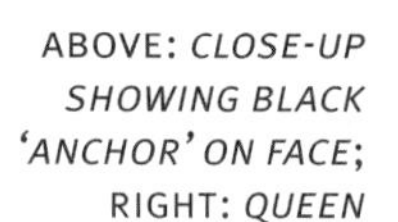

ABOVE: *CLOSE-UP SHOWING BLACK 'ANCHOR' ON FACE*; RIGHT: *QUEEN*

FACT FILE

ORDER Hymenoptera FAMILY Vespidae BODY LENGTH 11–19mm SIMILAR SPECIES There are several similar species, especially the German Wasp *V. germanica*. The Hornet (opposite) is much larger and browner. Some bees (pp. 150–152) and hoverflies (pp. 82–85) are similar.

JAN FEB MAR **APR MAY JUN JUL AUG SEP OCT** NOV DEC

HORNET

Vespa crabro

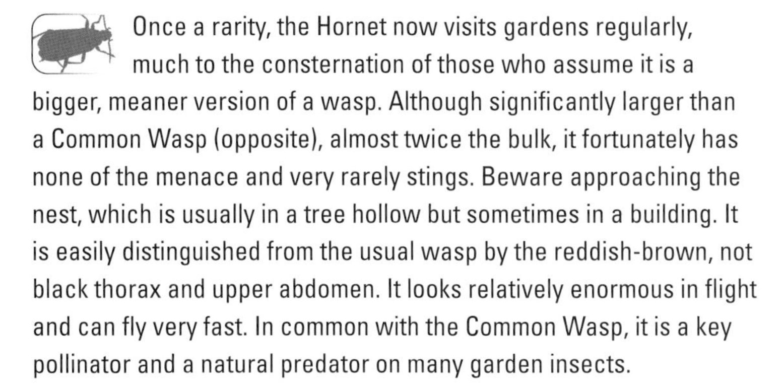

Once a rarity, the Hornet now visits gardens regularly, much to the consternation of those who assume it is a bigger, meaner version of a wasp. Although significantly larger than a Common Wasp (opposite), almost twice the bulk, it fortunately has none of the menace and very rarely stings. Beware approaching the nest, which is usually in a tree hollow but sometimes in a building. It is easily distinguished from the usual wasp by the reddish-brown, not black thorax and upper abdomen. It looks relatively enormous in flight and can fly very fast. In common with the Common Wasp, it is a key pollinator and a natural predator on many garden insects.

EGG (in nest of social colony) > **LARVA** (grub) > **PUPA** (in nest in colony) > **ADULT**. The larvae are fed by the workers on insect flesh. It overwinters as an adult queen.

A Hornet can fly 100km a day at 40kph.

WHERE TO FIND

Locally common over most of the region north to southern Scandinavia, but absent from Scotland and Ireland. It is expanding in many areas.

QUEEN NEST-BUILDING

FACT FILE

ORDER **Hymenoptera** FAMILY **Vespidae** BODY LENGTH **15–28mm** SIMILAR SPECIES **Other social wasps are much smaller. Larger than the largest hoverfly.**

JAN FEB MAR APR MAY **JUN JUL AUG SEP** OCT NOV DEC

ORNATE-TAILED DIGGER WASP

Cerceris rybyensis

WHERE TO FIND
Widespread and fairly common. In the UK, mainly southern.

This species spends much of its time visiting flowers for pollen. It also constructs a nest-hole by digging. The female seeks out solitary bees, paralyzes them with its sting and carries them back to its nest, held under the body. Each female brings four or five bees, which die and become food for the larvae.

EGG (in burrow) > **LARVA** (in burrow, feeds on bodies of prey species) > **PUPA** (cocoon) > **ADULT**.

FACT FILE

ORDER **Hymenoptera** FAMILY **Crabronidae** BODY LENGTH **6–12mm** SIMILAR SPECIES There are many similar, closely related species.

JAN FEB **MAR APR MAY JUN** JUL AUG SEP OCT NOV DEC

BUFFISH MINING BEE

Andrena nigroaenea

WHERE TO FIND
Locally common as far north as south Norway and Sweden. Rare in Scotland. Sparse but increasing in the Baltic, and localized in northern France.

These bees mine holes in lawns, earth and walls for their nests. They are common in urban areas with sometimes many hundreds of nests close together. Their pollen-carrying hairs are orange in colour, the thorax is buff coloured; the abdomen is dark.

EGG (in nest cell) > **LARVA** (grubs in cells) > **PUPA** (in nest) > **ADULT** Adults feed on nectar from a wide variety of flowers.

FACT FILE

ORDER **Hymenoptera** FAMILY **Apidae** BODY LENGTH **10–15mm** SIMILAR SPECIES About the same size as the Honey Bee (p. 152) , but with buff thorax and orange pollen basket. Similar to the Chocolate Mining Bee *A. scotica*, but this species does not have the hairy orange hindlegs.

JAN	FEB	MAR	**APR**	**MAY**	**JUN**	JUL	AUG	SEP	OCT	NOV	DEC

RED MASON BEE

Osmia bicornis

This is a solid, hairy and handsome bee that is very common in gardens, including urban ones. The female has remarkable small 'horns' on her head that are used for pattering down mud to help solidify it in the nest. The head is jet black, the thorax black and the abdomen a very attractive shade of brownish-red. This is a solitary bee that uses existing holes for its nest and is especially partial to bee hotels. It feeds on a variety of garden flowers, as well as blossom from fruit trees and other flowering crops.

EGG (up to 10, in chambers in nest, one per cell, the cell sealed with mud) > **LARVA** (grubs, hatch in autumn) > **PUPA** (in nest) > **ADULT**. Nests in pre-existing holes of many kinds – in cracks in walls and masonry, trees and similar places. Overwinters as a larva.

Urban mason bees sometimes use a keyhole for a nest. A single Red Mason Bee is equivalent to 120 worker Honey Bees (p. 152) in terms of its pollination service, since it spills so much pollen due to a lack of pollen baskets.

WHERE TO FIND

Common continental species found in the south of Scandinavia. In the British Isles, common in lowlands of England and Wales, even in urban areas; scarce in Scotland and absent from Ireland.

FACT FILE

ORDER Hymenoptera
FAMILY Apidae
BODY LENGTH 7–14mm
SIMILAR SPECIES Similar to many bee species.

ABOVE: *FEMALE*; LEFT: *MALE*

JAN FEB **MAR APR MAY JUN JUL AUG SEP OCT** NOV DEC

COMMON FURROW BEE

Lasioglossum calceatum

WHERE TO FIND
Common species throughout the region, often content in urban locations.

This is one of the 'under-the-radar' bees of the region. We have more than 300 species in all, and many are inconspicuous, like this one. It is small and compact and only looks like a bee when you take a close look and see its hairy thorax and pollen-carrying legs. This species is rather dark with a shiny abdomen with narrow stripes. The male sometimes has an orange tint to the abdomen.

EGG (in cell placed in nest chamber in side shoot from long, almost vertical tunnel mined by female) > **LARVA** (grubs in cells, tended by female, then a few workers) > **PUPA** (in nest) > **ADULT**. Mines burrow in loose soil.

Sometimes two or more females are found in a nest, in which case one becomes the dominant queen and the others workers.

FACT FILE

ORDER **Hymenoptera** FAMILY **Apidae** BODY LENGTH **8–11mm** SIMILAR SPECIES **One of many similar species.**

JAN	FEB	MAR	APR	**MAY**	**JUN**	**JUL**	**AUG**	SEP	OCT	NOV	DEC

PATCHWORK LEAFCUTTER BEE

Megachile centuncularis

The dark brown colours and the narrow pale rings on the black abdomen resemble those of the familiar Honey Bee (p. 152) at first sight – but then you notice the underside, which is startling orange. This dash of colour is the 'pollen brush', a series of hairs that carries the pollen while the bee is foraging. This is a solitary bee that builds a nest in a hole in a wall, or bee hotel. You are more likely to notice its work than the animal itself; it cuts very obvious, semi-circular holes in the leaves of roses, honeysuckles, birches and other plants, harvested for the nest. The bees forage especially on dandelions and thistles.

WHERE TO FIND

Common and widespread throughout most of the region south of central Scandinavia, but very patchy in Scotland and Ireland.

EGG (in nest cell with semi-liquid pollen and nectar, sealed with leaf) > **LARVA** (grubs in cells) > **PUPA** (in nest) > **ADULT**. Nests in pre-existing hole (in wall or tree), above the ground. Overwinters as a pupa.

★ The outer cells in the tubular nest always contain male offspring, the inner ones female offspring.

FACT FILE

ORDER Hymenoptera FAMILY Apidae BODY LENGTH 9–12mm SIMILAR SPECIES Honey Bee, but this species is distinguishable by its lack of pollen baskets.

JAN	FEB	MAR	**APR**	**MAY**	**JUN**	JUL	AUG	SEP	OCT	NOV	DEC

GOODEN'S NOMAD BEE

Nomada goodeniana

WHERE TO FIND

Common south of Scandinavia. In the British Isles, scarce in Scotland and Ireland.

The photo is not wrong – this is a bee, not a wasp. It is a nomad bee, an almost hairless type that lays its eggs in the nests of other bees, using its wasp-like pattern as protection and deception. Nomad bees are smaller than wasps, and this species has a much cleaner set of yellow-and-black bands across the abdomen, lacking any spots. There are more bands, and there is a narrow black waist at the front end of the abdomen. The species is common but easily overlooked; a good place to find one is on a flower finding nectar, or around the entrances of solitary bee nests.

EGG (laid into wall of unsealed cell inside nest of solitary bee, usually a mining bee, p. 146) > **LARVA** (grubs in cells kill host grubs with their sharp mandibles, and steal food provided for host grubs) > **PUPA** (in nest) > **ADULT**.

★ Some nomad bees sleep on plant stems and grasses, holding on by their mandibles.

NECTARING

FACT FILE

ORDER Hymenoptera FAMILY Apidae BODY LENGTH 9–13mm SIMILAR SPECIES There are lots of similar nomad bee species, and look out for many other wasps, which are similar.

HAIRY-FOOTED FLOWER-BEE

Anthophora plumipes

This solitary bee is a real garden character and a welcome sight as it is one of the earliest insects, let alone bees, to emerge in spring, sometimes by late February. This busy bee darts about energetically, with much hovering, often holding its tongue extended, feeding on primroses, grape hyacinths, lungwort and other early flowers. Most unusually for a bee, the male and female are quite different from each other. The female is black, hairy and bumblebee-like, with an orange tag of hairs on the hindlegs. The male is tawny-brown on the thorax with a darker abdomen and a whitish face, but check out the amazing middle legs – these have feathery extensions and are used to signal to females. The males also have quite characteristic miserable-looking faces.

EGG (in nest cell with semi-liquid pollen and nectar) > **LARVA** (grubs in cells) > **PUPA** (in nest) > **ADULT**. Nests in pre-existing hole (in wall, chimney or cliff), above the ground. The males emerge first, about two weeks before the females.

These bees have a habit of dropping down chimneys, looking for a place to nest.

WHERE TO FIND

Common as far north as southern Scandinavia and the Baltic States, but absent from Scotland and Ireland.

LEFT: *MALE*; RIGHT: *FEMALE*

FACT FILE

ORDER Hymenoptera FAMILY Apidae BODY LENGTH 14–17mm SIMILAR SPECIES Distinctive for its season and flight style. The female resembles a bumblebee but is smaller – the colouring of the hindlegs almost resembles pollen baskets, but solitary bees do not have pollen baskets as do bumblebees.

JAN | FEB | MAR | APR | MAY | JUN | JUL | AUG | SEP | OCT | NOV | DEC

HONEY BEE

Apis mellifera

WHERE TO FIND
Common throughout the region.

This is the bee everyone knows. It lives in huge colonies, makes honey in hives and can sting you. Most bees you see in gardens are from nearby domesticated colonies looked after by beekeepers. This bee (worker) is identified by its moderate size (much smaller than a bumblebee), dull brown thorax and dark abdomen with some pale yellow bands. It also has black legs. Drones (males) are larger and lack abdominal bands. The pollen sacs are obvious on workers. The hives are built of beeswax. Honey Bees live in colonies with a single queen, which spends its time laying eggs to replenish workers. The workers look after the larvae and pupae in the familiar honeycomb. The queen also produces drones and new queens. When a queen founds a new colony it goes on mating flights and copulates with a number of unrelated drones.

EGG (laid singly into hexagonal chamber in honeycomb within hive) > **LARVA** (grubs, lack legs and eyes, in chambers) > **PUPA** (in sealed chamber, hatches after a week) > **ADULT**.

WORKER

Dark abdomen with pale bands

Workers have tasks. When young they are colony cleaners, but as they get older they produce honeycomb, then become foragers. Some are professional undertakers, removing dead bodies. A returning foraging worker performs the famous 'waggle dance', its moves indicating the direction of good foraging areas and how far away they are. If it has to wait to give its message it will compensate for the sun's movement.

FACT FILE

ORDER Hymenoptera FAMILY Apidae BODY LENGTH 12–16mm SIMILAR SPECIES See solitary bees, and Common Drone Fly (p. 89).

JAN | FEB | MAR | APR | **MAY** | **JUN** | **JUL** | **AUG** | **SEP** | OCT | NOV | DEC

WOOL-CARDER BEE

Anthidium manicatum

This is a big, bullying bee with a unique appearance and unusual lifestyle. Its plump body is mainly dull brownish-black, but it sports a series of yellow spots on the sides of its abdomen that look like dashes of war paint. The amount of yellow varies, and so does the size of individuals, with some being much larger than others. This bee is associated with plants known as labiates (for example woundworts, especially lamb's ear). Males defend patches of the flowers and attack any bee of any species, or even any hoverfly, which intrudes, using the extraordinary sharp spines at the ends and sides of the abdomen as weapons. A male will hover, then fly straight at the offending rival and head butt or body charge it, sometimes fatally. A solitary bee, the female attends a single-hole nest.

WHERE TO FIND
Widespread but localized. Scarce in southern Scandinavia and seemingly absent from Ireland.

EGG (in chambers in nest, floats on mixture of honey and pollen) > **LARVA** (grub) > **PUPA** (in nest) > **ADULT**. Female visits plants with hairy surfaces and chews the wool off to line the cell walls, hence the common name of the species. Nests in pre-existing hole.

This is the only bee species in Britain in which the male is larger than the female. Males and females both mate with many individuals of the opposite sex.

RIGHT: *MALE*; TOP: *MALE, SHOWING VENTRAL SPINES*

FACT FILE

ORDER Hymenoptera FAMILY Apidae BODY LENGTH 9–17mm SIMILAR SPECIES An unusually distinctive bee. However, its markings are wasp-like, particularly in the female.

JAN	FEB	MAR	APR	MAY	JUN	JUL	**AUG**	**SEP**	**OCT**	**NOV**	DEC

IVY BEE

Colletes hederae

WHERE TO FIND
Found in the UK, Germany, Switzerland, France and the Netherlands. Not recorded in Britain until 2001.

This species looks like a cross between a bee and a wasp; the orange-brown and black striped abdomen is wasp-like, but the thorax is covered with hairs of a warm orange-buff. This is one of the latest bees in the year, the only one confined to autumn. It is a specialist on harvesting the pollen and nectar of ivy and is often seen mixing with late wasps and other bees at banks of ivy flowers, on sunny days well into November. This is a solitary bee, one of the 'plasterer bees', which makes a hole in loose, sandy soil on south-facing banks. Where conditions are suitable many hundreds may nest side by side. Males emerge first, females from mid-September. This can cause problems. Many males crowd to await emerging females and may form frenzied 'mating balls' of activity.

FEMALE

EGG (in chambers in hole nest, mined in soil by female) > **LARVA** (grub) > **PUPA** (in nest) > **ADULT**.

Remarkably, it was only described to science in 1993.

FACT FILE

ORDER **Hymenoptera** FAMILY **Apidae** BODY LENGTH **10–16mm** SIMILAR SPECIES **Slightly bigger than the Honey Bee (p. 152). The late season is unusual. It is distinguished from the Honey Bee by the lack of pollen baskets, brighter colouring of the orange hair on the thorax, and abdominal bands.**

JAN FEB **MAR APR MAY JUN JUL AUG SEP OCT** NOV DEC

COMMON CARDER BUMBLEBEE

Bombus pascuorum

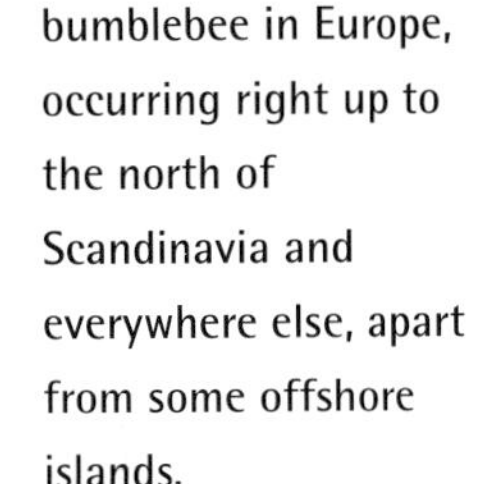

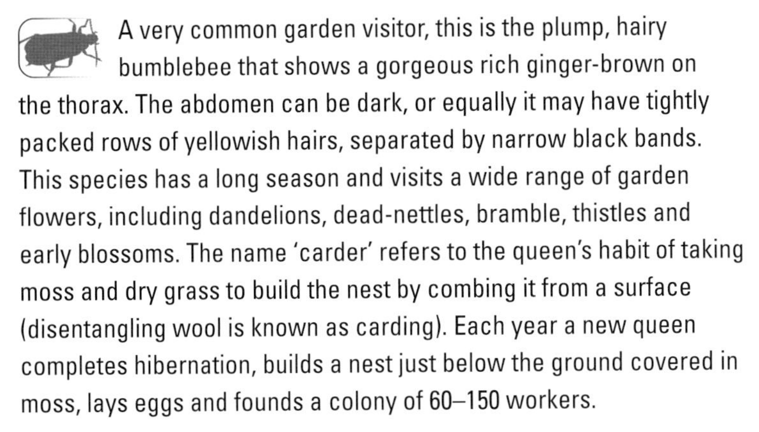

A very common garden visitor, this is the plump, hairy bumblebee that shows a gorgeous rich ginger-brown on the thorax. The abdomen can be dark, or equally it may have tightly packed rows of yellowish hairs, separated by narrow black bands. This species has a long season and visits a wide range of garden flowers, including dandelions, dead-nettles, bramble, thistles and early blossoms. The name 'carder' refers to the queen's habit of taking moss and dry grass to build the nest by combing it from a surface (disentangling wool is known as carding). Each year a new queen completes hibernation, builds a nest just below the ground covered in moss, lays eggs and founds a colony of 60–150 workers.

WHERE TO FIND

The most common bumblebee in Europe, occurring right up to the north of Scandinavia and everywhere else, apart from some offshore islands.

EGG (in chambers in nest) > **LARVA** (grubs in cells, fed by workers) > **PUPA** (in nest) > **ADULT**. Nests in pre-existing hole (of a mouse, for example), often among thick vegetation, usually but not always on the ground. Queens and all workers die off in autumn, while the offspring queens mate and hibernate.

The queen creates a 2cm-high cup within the nest filled with nectar, as a reserve against a rainy day.

FACT FILE

ORDER Hymenoptera FAMILY Apidae BODY LENGTH 10–13mm SIMILAR SPECIES Some less common bumblebees are of similar pattern and some smaller species, such as the Red Mason Bee (p. 147), are also reddish.

JAN FEB MAR APR MAY JUN JUL AUG SEP OCT NOV DEC

BUFF-TAILED BUMBLEBEE

Bombus terrestris

WHERE TO FIND
Common throughout the region to the southern third of Scandinavia and the Baltic States.

This is one of several types of big bumblebee whose queens appear very early in spring, from February onwards. The queen has an off-white or buff abdomen tip, but the later-appearing workers have whiter tips. It also has a narrow brownish collar behind the head and a pale, somewhat slapdash yellow band across the abdomen. The queens locate their nests underground, often in mouse holes, and at its peak the resulting colony may have 300 workers.

EGG (in chambers in nest) > **LARVA** (grubs in cells, first fed by queen, then by workers) > **PUPA** (in nest) > **ADULT**. Nests in pre-existing hole (that of a mouse, for example) underground. The queen mates in late summer, hibernates, wakes, and lays eggs in the nest, then feeds the larvae on pollen and nectar; they pupate and become workers.

Workers have an average foraging range of 660m. However, they can re-find the nest, even if it is 13km away. Increasingly, there are records of this bee being active in winter (November–February).

FACT FILE

ORDER Hymenoptera FAMILY Apidae BODY LENGTH 11–22mm SIMILAR SPECIES The very similar White-tailed Bumblebee *B. lucorum* queen has a white tail, and all castes have narrower, yellower and more sharply defined lateral bands. Buff-tailed queens have orange banding preceding the buff-to-white base of the abdomen. Workers of both are almost indistinguishable in the field, but the Buff-tailed's banding is more orange. Related species, such as the Polar Bee *B. polaris*, occur right up into the Arctic Circle.

JAN | FEB | MAR | **APR** | **MAY** | **JUN** | **JUL** | **AUG** | **SEP** | OCT | NOV | DEC

VESTAL CUCKOO BUMBLEBEE

Bombus vestalis

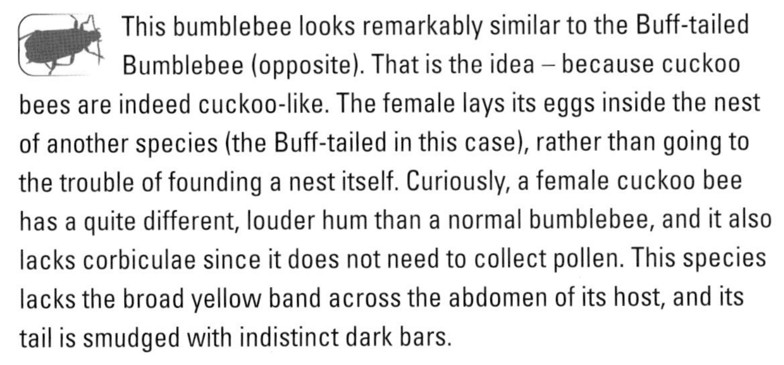

This bumblebee looks remarkably similar to the Buff-tailed Bumblebee (opposite). That is the idea – because cuckoo bees are indeed cuckoo-like. The female lays its eggs inside the nest of another species (the Buff-tailed in this case), rather than going to the trouble of founding a nest itself. Curiously, a female cuckoo bee has a quite different, louder hum than a normal bumblebee, and it also lacks corbiculae since it does not need to collect pollen. This species lacks the broad yellow band across the abdomen of its host, and its tail is smudged with indistinct dark bars.

EGG (laid inside active nest of host species) > **LARVA** (grubs in cells, looked after by host workers) > **PUPA** (in nest, same) > **ADULT** (queens first, males from May). The cuckoo bee queen mimics the appearance and smell of its host and enters its nest. Over time, she kills the older host workers (detected by their scent) one by one, and soon kills or evicts the host queen. Younger or unborn workers are retained for the purpose of looking after the cuckoo bee's next generation of queens and males.

Female cuckoo bees can find a host nest by following pheromone trails. If a colony is too large, the host workers can defend the nest from a cuckoo bee.

WHERE TO FIND

Fairly common in gardens in southern part of the region. Absent from all but extreme southern Scandinavia, from Denmark, Ireland and Scotland. Expanding into the Baltic States.

FACT FILE

ORDER Hymenoptera FAMILY Apidae BODY LENGTH 15–22mm SIMILAR SPECIES White-tailed Bumblebee *B. lucorum* and Buff-tailed Bumblebee (opposite).

FURTHER READING

Bibliography

Ball, S. & Morris, R. 2013. R. *Britain's Hoverflies.* WildGuides. Princeton University Press.

Brock, P. D. 2019. *A Comprehensive Guide to Insects of Britain and Ireland.* Pisces Publications.

Brock, P. D. 2021. *Britain's Insects.* WildGuides. Princeton University Press.

Chinery, M. 1993. *Field Guide to the Insects of Britain and Northern Europe.* Collins.

Falk, S. 2015. *Field Guide to the Bees of Great Britain and Ireland.* British Wildlife Field Guides. Bloomsbury.

Jones, R. 2018. *Beetles.* The New Naturalist Library. William Collins.

Labas et al. 2016. *Ants of Britain and Europe: A Photographic Guide.* Bloomsbury Wildlife.

Lewington, R. 2008. *Guide to Garden Wildlife.* British Wildlife Publishing.

McAlister, E. 2017. *The Secret Life of Flies.* Natural History Museum.

Marren, P. & Mabey, R. 2010. *Bugs Britannica.* Chatto & Windus.

Newland, D. et al. 2010. *Britain's Butterflies.* WildGuides. Princeton University Press.

Roy, H. & Brown, P. 2018. *Field Guide to the Ladybirds of Great Britain and Ireland.* Bloomsbury Wildlife Guides.

Smallshire, D. & Swash, A. 2014. *Britain's Dragonflies.* WildGuides. Princeton University Press.

Sterling, P. & Parsons, M. 2012. *Field Guide to the Micro-Moths of Great Britain and Ireland.* British Wildlife Publishing.

Waring, P. & Townsend, M. 2017. *Field Guide to the Moths of Great Britain and Ireland.* Bloomsbury Wildlife Guides.

Warren, M. 2021. *Butterflies: A Natural History.* British Wildlife Collection No. 10. Bloomsbury Wildlife.

Young, M. 1997. *The Natural History of Moths.* Poyser Natural History.

Websites

www.gailashton.co.uk

www.flickr.com/photos/63075200@N07/collections (mainly hoverflies)

www.britishbugs.org.uk

www.ukbeetles.co.uk

Societies

Amateur Entomologist's Society: www.amentsoc.org

British Dragonfly Society: https://british-dragonflies.org.uk

British Entomological and Natural History Society: www.benhs.org.uk

Bumblebee Conservation Trust: www.bumblebeeconservation.org

Butterfly Conservation: https://butterfly-conservation.org

Buglife (Invertebrate Conservation Trust): www.buglife.org

People's Trust for Endangered Species: https://ptes.org

Royal Entomological Society: www.royensoc.org

Reported Sightings

- Any insect sightings, including garden ones, can be reported to the Biological Records Centre at www.brc.ac.uk/recording-schemes.
- BeeWalk is a bumblebee monitoring scheme run by the Bumblebee Conservation Trust.
- Stag Beetle sightings should be reported to the People's Trust for Endangered Species.
- Sightings of dragonflies and damselflies can be entered at the website of the British Dragonfly Society.
- Butterfly Conservation runs butterfly and moth surveys and you can submit records from your garden and elsewhere to:

Butterflies for the New Millennium,
Garden Butterfly Survey,
National Moth Recording Scheme.

INDEX

Tick boxes are included next to the English name of each species so you can mark off species that you have seen.

ACKNOWLEDGEMENTS

The authors wish to extend grateful thanks to John Beaufoy for believing in us and enabling us to produce this book. We are also grateful to Rosemary Wilkinson for managing the project with such proficiency, patience and grace. Krystyna Mayer has done a wonderful job in editing and we also thank Nigel Partridge for bringing the design to life.

Dominic Couzens would like to thank, as ever, his family for coping with him during the torturous process of writing a book: thanks to Carolyn, Emmie and Sam.

Gail Ashton would like to thank Ross Piper, for igniting her love for insects and patiently explaining the point of wasps all those years ago; the Herts Invertebrate Project, in particular Dan Asaw, William Bishop, Ian Carle, Joe Gray and Simon Knott, for welcoming her into their world and sharing their extraordinary knowledge; DC, for taking a huge gamble on a rookie; her family, for supporting her lifelong love of nature and photography and, most of all, Martha & Zac, who put up with mum's crazy projects and a house full of invertebrates – love you both.